Green Building and LEED Core Concepts Guide

SECOND EDITION

Purchase agreement and license to use
GREEN BUILDING AND LEED® CORE CONCEPTS GUIDE, Second Edition

The U.S. Green Building Council (USGBC) devoted significant time and resources to create this Guide and all of its LEED™ publications. All LEED publications are protected by statutory copyright and trademark protection within the United States and abroad. Your possession of The Green Building and LEED Core Concepts Guide, Second Edition (the "Guide"), constitutes ownership of a material object and in no way constitutes a conveyance of ownership or entitlement to copyrighted materials contained herein. As a result, you are prohibited by law from engaging in conduct that would constitute infringement upon the exclusive rights retained by USGBC.

USGBC authorizes individual, limited use of the Guide, subject to the terms and conditions contained herein. In exchange for this limited authorization, the user agrees:

(1) to retain all copyright and other proprietary notices contained in the Guide;

(2) not to sell or modify any copy of the Guide in any way; and

(3) not to reproduce, display or distribute the Guide in any way for any public or commercial purpose, including display on a website or in a networked environment.

Unauthorized use of the Guide violates copyright, trademark, and other laws and is prohibited.

The text of the federal and state codes, regulations, voluntary standards, etc., reproduced in the Guide is used under license to USGBC or, in some instance, in the public domain. All other text, graphics, layout and other elements of content in the Guide are owned by USGBC and are protected by copyright under both United States and foreign laws.

NOTE: FOR DOWNLOADS OF THE GUIDE:

Redistributing the Guide on the internet, in any other networked environment, in any digital format, or otherwise is STRICTLY prohibited, even if offered free of charge. DOWNLOADS OF THE GUIDE MAY NOT BE COPIED OR DISTRIBUTED. THE USER OF THE GUIDE MAY NOT ALTER, REDISTRIBUTE, UPLOAD OR PUBLISH THIS GUIDE IN WHOLE OR IN PART, AND HAS NO RIGHTS TO LEND OR SELL ANY COPY OF THE DOWNLOAD TO OTHER PERSONS. DOING SO WILL VIOLATE THE COPYRIGHT OF THE GUIDE.

Disclaimer

None of the parties involved in the funding or creation of the Guide, including the U.S. Green Building Council (USGBC), its members, contractors, affiliates or the United States government, assume any liability or responsibility to the user or any third parties for the accuracy, completeness, or use of or reliance on any information contained in The Guide. The Guide is not associated with, nor endorsed by the Green Building Certification Institute (GBCI) and does not guarantee a successful outcome on any examination mentioned herein or associated with GBCI or USGBC. Although the information contained in the Guide is believed to be reliable and accurate, the Guide is provided as-is with no

Published by:

U.S. Green Building Council
2101 L Street NW
Suite 500
Washington, DC 20037

Trademark

LEED® is a registered trademark of the U.S. Green Building Council.
ISBN: 978-1-932444-50-6

Project Team
CTG Energetics, Inc.

Heather Joy Rosenberg, Principal

Karen Blust, Green Building Consultant

Natalie Bodenhamer, Green Building Consultant

Clare Jones, Green Building Analyst

Lani Kalemba, Green Building Consultant

Joshua Joy Kamensky, Consultant

Joel Todd, Environmental Consultant

Second Edition Guide Review Team

John Boecker, 7group

Nick Rajkovich, University of Michigan

Kathy Roper, Georgia Institute of Technology

Chris Shaffner, The Green Engineer, LLP

Lynn Simon, Simon & Associates, Inc.

USGBC Staff

Julia Feder, Director of Educational Technology

Karol Kaiser, Director of Education Development

Jenny Poole, Manager of Education Media

Jen Schill, Manager of LEED Education Development

Jacquelyn Erdman, Knowledge Center Coordinator

Jacob Monroe, Education Resources Coordinator

CONTENTS

IMAGINE IT

Imagine getting up on a warm spring morning and deciding it's the perfect day to ride your bike to work. Invigorated by your morning ride and eager to start the day, you head into your office. As you pass through a common area, you see a group of coworkers deep in a collaborative work session. They're seated around a gorgeous oak table hand-crafted by local artisans and made entirely of wood reclaimed from a tree that fell naturally in a nearby forest.

Imagine getting to your desk and sitting down without flipping a light switch—the huge floor-to-ceiling windows nearby provide plenty of natural springtime light, and if it gets cloudy this afternoon, sensors in your work area will kick on overhead lighting to an appropriate level of brightness. Meanwhile, your personal control of the temperature in your work area allows you to stay warm even as your neighbor, who has a higher cold tolerance, works at a temperature that's comfortable for him.

Imagine being surrounded by decorative elements that invoke nature and keep you connected to the natural world even while you're inside. Imagine an herb garden in the office cafeteria and an educational display in the office lobby—constant reminders for you and your company's visitors of just what it is that makes your building so special.

And imagine leaving the office to find that it has started raining. But not to worry, you just duck around the corner to one of the many bus stops nearby. You mount your bike to the rack on the front of the bus and climb aboard.

You settle into your seat at the end of a full day of work, feeling the positive effects of having spent your day in an environment filled with clean indoor air, with plenty of exposure to natural light. Your mind is clear and your energy and spirits high, knowing that your workday cost substantially less in energy and water use than it would have in a more traditional building.

This is what it feels like for me and my colleagues at the LEED Platinum U.S. Green Building Council headquarters in Washington, D.C. It is what it's like for the thousands upon thousands of people worldwide who work in LEED-certified office space. And if you tweak the details, it is what it's like for all the students nationwide who study in green schools and live in green dorms, and for the increasing number of families who live in green homes.

Now, imagine that designing, building, operating, marketing, supporting, or celebrating green buildings was at the heart of your everyday work. Imagine being a green building professional.

With the *Green Building and LEED Core Concepts Guide*, you're on your way to just such a career. We hope you enjoy the journey, and we look forward to the innovations you'll bring as part of the green building community.

Rick Fedrizzi
President, CEO and Founding Chair
U.S. Green Building Council

SECTION 1
INTRODUCTION TO GREEN
BUILDINGS AND COMMUNITIES

Our built environment is all around us; it provides the setting for all our lives' events, big and small. And whether we notice it or not, our built environment plays a huge role in our natural environment, our economic environment, and our cultural environment. The built environment provides a context for facing and addressing humankind's greatest contemporary challenges.

Green building is fundamentally a process of continual improvement. It is a process by which today's "best practices" become tomorrow's standard practices, a rising foundation for ever-higher levels of performance. Green building can help us create more vital communities, more healthful indoor and outdoor spaces, and stronger connections to nature. The green building movement strives to effect a permanent shift in prevailing design, planning, construction, and operations practices, resulting in lower-impact, more sustainable, and ultimately regenerative built environments.

For the purposes of this guide, "built environment" refers to any environment that is man-made and provides a structure for human activity. These environments range from shelters and individual buildings to neighborhoods and vast metropolitan areas. This guide explains the reasons we must change traditional building practices. It presents fundamental concepts of green building and provides a summary of the application strategies that will help you be a more effective participant in the green building process.

The remainder of this section of the guide gives the rationale for green building and the related concept of sustainability. The core concepts of sustainable thinking are explored in Section 2. Section 3 looks at important components of the sustainable design and operations process. Section 4 reviews the application of green technologies and strategies. Section 5 offers more information on the programs of the U.S. Green Building Council (USGBC), particularly the Leadership in Energy and Environmental Design (LEED) certification system. Additional resources are listed in the Appendix, and educational opportunities to support your growth and success as a green building professional are available from USGBC at usgbc. org/education.

THE ENVIRONMENTAL IMPACTS OF BUILDINGS

Why is green building necessary? Buildings and communities, including the resources used to create them and the energy, water, and materials needed to operate them, have a significant effect on the environment and human health. In the United States, buildings account for:

- 14% of potable water consumption[1]
- 30% of waste output
- 40% of raw materials use[2]
- 38% of carbon dioxide emissions
- 24% to 50% of energy use
- 72% of electricity consumption[3]

> **The cumulative effect of conventional practices in the building industry has profound implications for human health, the environment, and the economy:**
>
> - Clearing of land for development often destroys wildlife habitat
> - Extracting, manufacturing, and transporting materials may pollute water and air, release toxic chemicals, and emit greenhouse gases
> - Building operations require large inputs of energy and water and generate substantial waste streams
> - Transportation to and from buildings by commuters and service providers compounds the harmful environmental effects associated with vehicle use, such as increased energy consumption and pollution

By building green, we can reduce that environmental damage. In many cases, green buildings can even enhance the health of the environment and the people who use them.

A study by the New Buildings Institute found that in green buildings, average energy use intensities (energy consumed per unit of floor space) are 24% lower than in typical

---

1   J.F. Kenny, N.L. Barber, S.S. Hutson, K.S. Linsey, J.K. Lovelace, & M.A. Maupin. Estimated use of water in the United States in 2005: U.S. Geological Survey Circular 1344, (2009).
2   D.M. Roodman & N. Lenssen "A Building Revolution: How Ecology and Health Concerns Are Transforming Construction," Worldwatch Paper 124 (Worldwatch Institute, 1995).
3   Energy Information Administration, *EIA Annual Energy Outlook* (EIA, 2008).

buildings.[4] Additionally, the U.S. General Services Administration surveyed 12 green buildings in its portfolio and found these savings and improvements:

- 26% less energy usage
- 27% higher levels of occupant satisfaction
- 13% lower maintenance costs
- 33% lower emissions of carbon dioxide (CO_2)[5]

The study concluded that the federal government's green buildings outperform national averages in all measured performance areas—energy, operating costs, water use, occupant satisfaction,

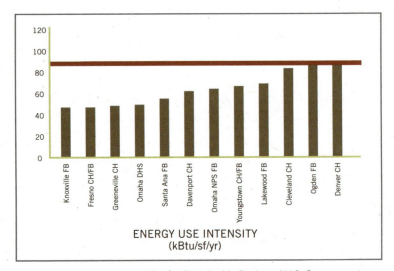

Figure 1.1. Energy Use Intensities for Sustainably Designed U.S. Government Buildings (Source: GSA 2008)
The red bar indicates the national average energy use intensity.

and carbon emissions. The agency attributed this performance to a fully integrated approach to sustainable design that addressed environmental, financial, and occupant satisfaction issues. This higher performance will last throughout a building's lifetime if the facility is also operated and maintained for sustainability.

WHAT IS GREEN BUILDING?

Sustainability is not a one-time treatment or product. Instead, green building is a process that applies to buildings, their sites, their interiors, their operations, and the communities in which they are situated. The process of green building flows throughout the entire life cycle of a project, beginning at the inception of a project idea and continuing seamlessly until the project reaches the end of its life and its parts are recycled or reused.

In this guide, the term **green building** encompasses planning, design, construction, operations, and ultimately end-of-life recycling or renewal of structures. Green building pursues solutions that represent a healthy and dynamic balance between environmental, social, and economic benefits.

Sustainability and "green," often used interchangeably, are about more than just reducing environmental impacts. Sustainability means creating places that are environmentally responsible, healthful, just, equitable, and profitable. Greening the built environment means looking holistically at natural, human, and economic systems and finding solutions that support quality of life for all.

---

4   Turner, C. & Frankel, Energy Performance of LEED® for New Construction Buildings (2008), http://www.newbuildings. org/sites/default/files/Energy_Performance_of_LEED-NC_Buildings-Final_3-4-08b.pdf.

5   Public Buildings Service, "Assessing Green Building Performance: A Post Occupancy Evaluation of 12 GSA Buildings" (General Services Administration, 2008), http://www.gsa.gov/graphics/pbs/GSA_Assessing_Green_Full_Report.pdf.

Triple bottom line is also often used to refer to the concept of sustainability. The term was coined by John Elkington, cofounder of the business consultancy SustainAbility, in his 1998 book *Cannibals with Forks: the Triple Bottom Line of 21st Century Business*. First applied to socially responsible business, the term can characterize all kinds of projects in the built environment. The triple bottom line concept incorporates a long-term view for assessing potential effects and best practices for three kinds of resources:

● **People (social capital).** All the costs and benefits to the people who design, construct, live in, work in, and constitute the local community and are influenced, directly or indirectly, by a project

● **Planet (natural capital).** All the costs and benefits of a project on the natural environment, locally and globally

● **Profit (economic capital).** All the economic costs and benefits of a project for all the stakeholders (not just the project owner)

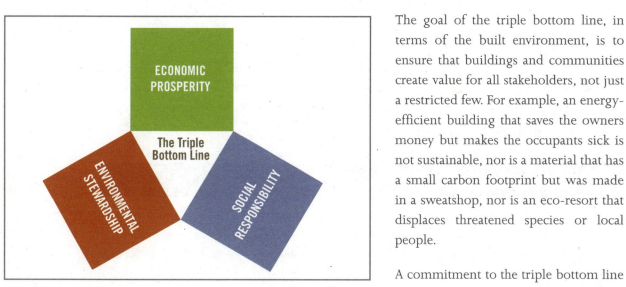

Figure 1.2. The Triple Bottom Line

The goal of the triple bottom line, in terms of the built environment, is to ensure that buildings and communities create value for all stakeholders, not just a restricted few. For example, an energy-efficient building that saves the owners money but makes the occupants sick is not sustainable, nor is a material that has a small carbon footprint but was made in a sweatshop, nor is an eco-resort that displaces threatened species or local people.

A commitment to the triple bottom line means a commitment to look beyond the status quo. It requires consideration of whole communities and whole systems, both at home and around the world. Research is needed to determine the impacts of a given project and find new solutions that are truly sustainable. New tools and processes are required to help projects arrive at integrative, synergistic, sustainable solutions.

The triple bottom line requires a shift in perspective about both the costs and the benefits of our decisions. The term **externalities** is used by economists to describe costs or benefits incurred by parties who are not part of a transaction. For example, the purchase price of a car does not account for the wear and tear it will have on public roads or the pollution it will put into the environment. To shift the valuation process to account for such negative externalities, building professionals require new metrics. The green building process and rating systems have begun to encourage quantification of externalities. The focus has been first on environmental metrics, but the list is expanding to include indicators of social justice and public health.

Making buildings more healthful, more comfortable, and more conducive to productivity for their occupants has special significance in light of studies conducted by the U.S. Environmental Protection Agency (EPA), which found that people in the United States spend, on average, 90% of their time indoors.[6] Occupants of green buildings are typically exposed to far lower levels of indoor pollutants and have significantly greater satisfaction with air quality and lighting than occupants of conventional buildings. Research conducted at Carnegie Mellon University shows that these benefits can translate into a 2% to 16% increase in workers' and students' productivity. Even small increases in productivity can dramatically increase the value of a building.[7]

THE RISE OF THE GREEN BUILDING INDUSTRY

Many of the elements of green building are not new or even unique. Before the widespread availability of inexpensive fossil fuels for energy use and transportation, builders understood the principles of **passive design**, capturing sunlight and wind for natural lighting, heating, and cooling. In many ways, green building represents a return to simpler, low-tech solutions. At the same time, there are now many high-tech strategies available to improve the performance of the built environment. Green building is about finding the best combination of solutions to create built environments that seamlessly integrate the best of the old and the new in intelligent and creative ways.

The USGBC was formed in 1992, a time when the field was beginning to define itself, to promote and encourage green building. A member-based organization, the USGBC community engages hundreds of thousands of individuals. The mission of USGBC is "to transform the way buildings and communities are designed, built and operated, enabling an environmentally and socially responsible, healthy, and prosperous environment that improves the quality of life."[8] USGBC supports achievement of this mission through education programs, advocacy, research, an extensive network of local chapters, and the **Leadership in Energy and Environmental Design (LEED) rating system**.

The Chesapeake Bay Foundation, an environmental advocacy, restoration, and education organization, is headquartered in Annapolis, Maryland.
Photo credit: Robb Williamson

Soon after it was formed, USGBC began developing LEED for rating and certifying the sustainability of buildings in the United States. Experts identified characteristics and performance levels that contributed to a definition of a green building. The first LEED green building rating system was launched in 1999. In the decade that followed, LEED expanded to include systems to rate the entire life cycle of the built environment, including land-use

6   U.S. Environmental Protection Agency, Report to Congress on Indoor Air Quality, volume 2, EPA/400/1-89/001C (EPA, 1989).
7   V. Loftness, V. Hartkopf, B. Gurtekin, and Y. Hua, "Building Investment Decision Support (BIDS™): Cost-Benefit Tool to Promote High Performance Components, Flexible Infrastructures and Systems Integration for Sustainable Commercial Buildings and Productive Organizations," Report on university research (AIA, 2005).
8   U.S Green Building Council, Strategic Plan 2009–2013 (USGBC, 2008).

planning and design-through-operations. It now provides rating systems for a wide array of building types, such as offices, schools, retail establishments, homes, and neighborhoods.

The trend toward green building practices in the United States has quickened in the past decade, contributing to a transformation in the market of building products and services, as well as the demand for skilled professionals. As more green products and technologies become available, the more mainstream green building will become.

Federal, state, and local governments are among those adopting sustainable building practices and policies. For example, the U.S. General Services Administration requires that all new federal government construction projects and substantial renovations achieve certification under the LEED rating system, and it encourages projects to achieve at least Silver certification.[9] Government agencies, utility companies, and manufacturers increasingly offer financial incentives for developers and owners to enhance the environmental performance of their buildings. The goal of LEED is market transformation—to fundamentally change how we design, build, and operate buildings and communities—through certification that honors levels of achievement in areas such as energy savings, water efficiency, CO_2 emissions reduction, improved indoor environmental quality, and stewardship of resources.

LEED applies to a wide range of commercial building types as well as residential structures. It addresses the complete building life cycle, from design and construction to operations and maintenance, tenant fitout, and significant retrofit. LEED for Neighborhood Development extends the benefits of green building beyond the footprint of a structure and into the broader community it serves. More information on USGBC and LEED is provided in Section 5.

GREEN BUILDING AND CLIMATE CHANGE

Although many environmental impacts are associated with buildings and addressed by rating systems such as LEED, climate change deserves special consideration because buildings and land-use are responsible for a large proportion of greenhouse gas emissions. To be effective, the policies that are emerging at the local, state, and federal levels to regulate greenhouse gas emissions must reflect a clear understanding of the connection between climate change and the built environment. Unfortunately, it is not enough for green building to lessen the effects that humans have on our climate. It must also prepare us for the inevitable consequences of climate change on our homes, communities, and society as a

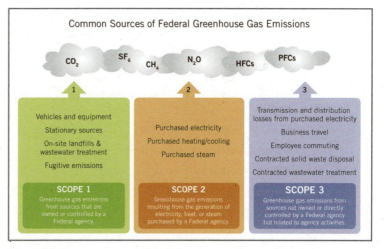

Figure 1.3. Common sources of greenhouse gas emissions from Federal facilities as called out by Executive Order 13514.

---

9   U.S. General Services Administration, LEED Building Information (GSA, 2010), http://www.gsa.gov.

whole. A lower-carbon future will not only have higher-performing buildings but also require higher-performing communities.

The built environment, including buildings and transportation systems, accounts for more than two-thirds of all greenhouse gas emissions.[10] Greenhouse gas emissions come from many components of the built environment, including building systems and energy use, transportation, water use and treatment, land-cover change, materials, and construction. By improving the efficiency of buildings and communities, we can significantly reduce greenhouse gas emissions.

However, focusing on building design and construction alone will not achieve the emissions reduction that scientists believe is required to mitigate climate change. Building location is equally important. For example, a typical code-compliant 135,000-square-foot office building in a suburban location will be responsible for approximately 8,375 tons (T) of carbon, or 11.8 T per person. Because this building is in the suburbs, emissions from transportation—people driving to and from work—make up half the total emissions associated with the project.

When that same building is moved to a location that is accessible via public transportation, bicycling, or walking, its total emissions decrease. The emissions from transportation are much less, and the relative amount from the building systems increases.

Figure 1.4. Building location without supporting infrastructure and services

When the building is designed and maintained as a green building with improved energy and water performance, the total emissions fall to 3,233 T, or 4.6 T per person. This example demonstrates the important link between buildings and land use and the need to address both to achieve meaningful reductions in greenhouse gas emissions.

Carbon emissions provide a useful metric for many aspects of green buildings and communities, including energy, water, solid waste, materials, and transportation, but green building

Figure 1.5. Building location with infrastructure and services

10   Energy Information Administration, *Annual Energy Outlook 2008* (EIA, 2008), http://www.eia.doe.gov/oiaf/aeo/pdf/0383(2008).pdf.

involves more than reducing greenhouse gas emissions. It is important to set goals for other issues as well, such as indoor air quality, healthy communities, and habitat protection. This comprehensive goal-setting process encourages programs and policies that will lead to sustainable communities. The goal-setting process will be discussed in Section 3.

ENERGY CONSUMPTION: BUILDING-ASSOCIATED TRANSPORTATION VERSUS OPERATIONS

For an average office building in the United States, 30 percent more energy is expended by office workers commuting to and from the building than is consumed by the building itself for heating, cooling, lighting, and other energy uses. Even for an office building built to modern energy codes (ASHRAE 90.1–2004), more than twice as much energy is used by commuters than by the building.[11]

Flexibility and adaptability are increasingly important attributes of green projects. Although the long-term effects of climate change are uncertain, we know that sea levels will be higher, temperatures higher, droughts longer and more widespread, and flooding more intense. How different regions will experience these changes will vary considerably, and building professionals will have to assess the likely threats to their communities and respond accordingly.

GREEN BUILDING OVER TIME

Green projects must be prepared to adapt to future change and be designed and operated to stand the test of time. Continuous monitoring is required to identify needed improvements and users' changing needs. Project teams must look far ahead to determine what stressors a project is likely to encounter and then build resilience into the system.

For example, where water supply depends on local snowpack, planning and design efforts might focus on water conservation, water storage, and alternative sources of water in anticipation that the snowpack will shrink. Where summer heat is already high, green builders will have to consider what will happen with even hotter temperatures and ensure that the cooling strategies of buildings can handle higher degree-days and still maintain air quality, which will be exacerbated at higher temperatures. These strategies and others will be discussed in Section 4.

The performance of most systems degrades with time, and thus a building's total emissions footprint incrementally increases over time unless care is taken to maintain the systems properly. Figure 1.6 illustrates building performance by looking at the total amount of carbon emissions over a building's life cycle.

11   H. Levin. Driving to Green Buildings: *The Transportation Energy Intensity of Buildings. Environmental Building News*, 16:9 (2007). http://www.buildinggreen.com

Building commissioning helps project teams ensure that systems are designed efficiently, are installed appropriately, and operate as intended. Commissioning is the process of verifying and documenting that a building and all its systems and assemblies are planned, designed, installed, tested, operated, and maintained to meet the owner's project requirements. However, even if initial performance is optimal, emissions will rise as performance falls over time. This trend can be periodically reversed through **retrocommissioning**, a tune-up that identifies inefficiencies and restores

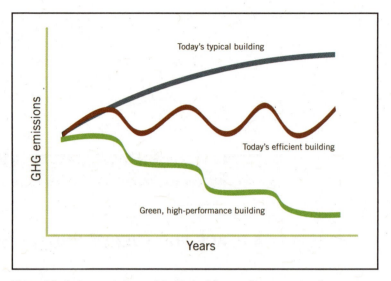

Figure 1.6. Carbon emissions related to building performance over time

high levels of performance. Commissioning and retrocommissioning will be reviewed in further detail in Section 4.

Green building professionals have a goal of following and achieving a path of continuous improvement. Because projects must be designed for the future, their operators need to participate in the design process and obtain the information they will need to monitor and maintain the building's performance. Monitoring and verification systems enable facilities personnel to identify and resolve issues that arise over time and even enhance a building's performance throughout the life of the project.

A chief goal of green building practitioners is to find new uses for existing structures. **Adaptive Reuse** means designing and building a structure in a way that makes it suitable for a future use different than its original use. Buildings can also be designed to prevent future obsolescence; for example, a flexible floor plan can accommodate offices today and apartments tomorrow. This avoids the environmental consequences of extracting materials for a new building and disposing of demolition waste. The adapted building reuses a site that is already served by infrastructure and avoids the conversion of farmland or forest to development. Designing a project to meet both current and evolving needs is one key to sustainability.

Adaptability is also critical for land use and municipal infrastructure, such as roads. Once road networks are established, they can remain fixed for centuries. In Rome, for example, the roadways that existed in ancient times have become today's automobile roads. This issue is particularly important as we move toward a lower-carbon future. Alternative transportation, including availability of public transportation, is essential for reducing carbon emissions. However, options for alternative and public transit, including bicycling and walking, depend on the proximity of destinations, connectivity of the community, and design of surroundings. Roads that are designed for only motor vehicles do not provide the flexibility or adaptability of a transportation network designed for diverse travel modes.

Buildings that protect the history and character of a place also promote sustainability. A project team can take advantage of the community's past by reusing materials with historic value. Linking the present with the past reinforces a sense of place and helps create attractive communities with viable commercial centers, discouraging sprawl. Sustainable design ensures that buildings and communities will survive and thrive for generations, no matter what the future holds.

GREEN BUILDING AND LOCATION

> ### A PLACE FOR EVERYTHING, EVERYTHING IN ITS PLACE.
>
> Benjamin Franklin

Location is a critical element of green building: it can define appropriate strategies, yet it can also limit how green a project can actually be. Depending on the environmental issues that are most critical in a particular area, location can influence a project team's priorities. Location includes these factors:

- **Natural context.** Climate, sun, wind, orientation, soils, precipitation, local flora and fauna
- **Infrastructural context.** Available resources, materials, skills, and connections to utilities, roads and transit
- **Social context.** Connections to the community and other destinations, local priorities, cultural history and traditions, local regulations and incentives

Selecting a location is one of the earliest decisions made in a project, and this decision defines many of the opportunities and constraints that the project team will encounter. It can determine whether a project can take advantage of sunlight, have access to public transportation and other services, and protect habitats. As discussed earlier in this section, a building whose occupants must drive long distances may contribute to greenhouse gas emissions, as well as destruction of natural habitat for infrastructure development.

To design sustainably for place, a team can start with a project site and determine what uses are most appropriate there. Alternatively, the team can start with a function and find the best place to put it. In either case, the goals of the project must be clear and the needs and resources must be clearly identified so that the building can be carefully integrated into its context and support a thriving and sustainable local community.

Project teams with a goal of sustainability develop a deep understanding of the place and context in which their projects are built. They go beyond a cursory site assessment and study the land and its history. They look for ways to make connections to the immediate site, the surrounding watershed, or ecological features and promote their healthy evolution. They also engage the community's traditions, strengths, and needs in order to ascertain how the project can contribute to social and economic well-being and growth.

Project Case Study

PRAIRIE CROSSING

For the Prairie Crossing development, a "sense of place" guides decision making. Located approximately 45 miles north of Chicago, developers knew they needed to plan for easy commuting access to the metropolitan region. Two Metra, northeast Illinois's commuter rail system, stations are within walking distance of this mixed use neighborhood. These rail lines help residents get to Chicago's O'Hare International Airport in 35 minutes and downtown Chicago in less than an hour. Additionally, cultivating a sense of place closer to home is at the forefront of developers land use planning efforts. Over 60 percent of the 677 acre development is legally protected land for wildlife and active use. The developers placed a high value on biking and walking and included over ten miles of such trails throughout the development. All of these characteristics culminated in Prairie Crossing being a LEED Certified Neighborhood Development project in the ND pilot. To learn more about Prairie Crossing, visit http://www.prairiecrossing.com.

GREEN BUILDING COSTS AND SAVINGS

At first glance, the additional work and alternative materials needed to build green may seem like a burdensome cost, but closer attention reveals this perception to be misleading. If sustainability is viewed as an expensive add-on to a building, we would mistake efforts to reduce energy costs or improve indoor environmental quality as comparable to specifying a better grade of countertop or a more impressive front door. Under this approach, any improvement beyond a minimally code-compliant baseline looks like an added cost.

If, however, we consider energy improvements part of an overall process, we often find that the added costs are balanced by savings over time. The initial expenditures continue to pay back over time, like a good investment. The best returns on these investments are realized when green building is integrated into the process at the earliest stages rather than as a last-minute effort. For instance, specification of more costly, high-performance windows may allow for the use of a smaller, lower-cost heating, ventilation, and air-conditioning (HVAC) system. More fundamentally, if we view sustainable design as part of the necessary functional requirements for building an energy-efficient structure and providing a safe, healthful environment, we can compare the cost of the green building with that of other buildings in the same class, rather than against an artificially low baseline.

A landmark study by the firm Davis Langdon found no significant difference between the average cost of a LEED-certified building and other new construction in the same category: there are expensive green buildings, and there are expensive conventional buildings. Certification as a green building was not a significant indicator of construction cost.[12]

Interestingly, the public dramatically overestimates the marginal cost of green building. A 2007 public opinion survey conducted by the World Business Council for Sustainable Development found that respondents believed, on average, that green features added 17% to the cost of a building, whereas a study of 146 green buildings found an actual average marginal cost of less than 2%.[13]

Green building is, however, a significant predictor of tangible improvements in building performance, and those improvements have considerable value. Studies have shown that certified green buildings command significantly higher rents. A University of California–Berkeley study analyzed 694 certified green buildings and compared them with 7,489 other office buildings, each located within a quarter-mile of a green building in the sample. The researchers found that, on average, certified green office buildings rented for 2% more than comparable nearby buildings. After adjusting for occupancy levels, they identified a 6% premium for certified buildings. The researchers calculated that at prevailing capitalization rates, this adds more than $5 million to the market value of each property.[14]

---

12 L.F. Matthiessen and P. Morris, "Cost of Green Revisited: Reexamining the Feasibility and Cost Impact of Sustainable Design in the Light of Increased Market Adoption" (Davis Langdon, 2007), http://www.davislangdon.com.

13 G. Kats et al., *Green Buildings and Communities: Costs and Benefits* (Good Energies, 2008).

14 P. Eichholtz, N. Kok, and J.M. Quigley, "Doing Well by Doing Good? Green Office Buildings" (Institute of Business and Economic Research, University of California–Berkeley, 2008), http://www.mistra.org/download/18.39aa239f11a8dd8de6b800026477/IBER+Green+Office+Buildings+NKok+et+al.pdf.

BEYOND GREEN

Initially, green buildings were intended to reduce the damage to the environment and human health caused by creating and maintaining buildings and neighborhoods. As the concept of sustainability was applied to the built environment, it has become clear that doing less damage is not enough.

Leaders in the field now speak about buildings and communities that are **regenerative**, meaning that these sustainable environments evolve with living systems and contribute to the long-term renewal of resources and life. Some practitioners have begun to explore what it would mean to move beyond "sustainable" and participate as a positive developmental force in our ecosystems and communities. The focus is on building a comprehensive understanding of the place in which the project is located, recognizing the site's patterns and flow of life. Accordingly, such projects contribute to the healthy coevolution of humans and all life in that place. They thrive on diversity, for example, and clean the air rather than pollute it. Regenerative projects and communities involve stakeholders and require interactivity.

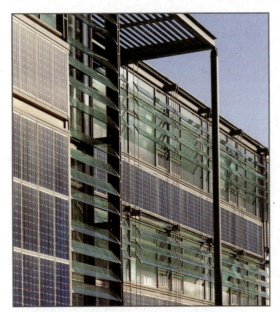

Figure 1.7. Regenerative Design

Regenerative projects support the health of the local community and regional ecosystems, generate electricity and send the excess to the grid, return water to the hydrologic system cleaner than it was before use, serve as locations for food production and community networking, regenerate biodiversity, and promote many other relationships that link the projects to the whole system of life around them.

Regenerative projects strive toward "net-zero"—using no more resources than they can produce. For example, **net-zero energy** projects use no more energy from the grid than they generate on site. These projects may be connected to the grid, drawing electricity from it at night and contributing energy from on-site renewable energy systems during the day, such that their total energy cost is zero. Other projects strive for **carbon neutrality**, emitting no more carbon emissions than they can either sequester or offset. Still other projects are designed to achieve a more even **water balance**: they use no more water than that which falls on site as precipitation, or they produce zero waste by recycling, reusing, or composting all materials.

Not all projects can achieve those levels of performance. Nevertheless, on average, green buildings save energy, use less water, generate less waste, and provide more healthful, more comfortable indoor environments. Specific strategies will be discussed in Section 4 of this guide.

Getting to green and beyond requires more than learning about new technologies and strategies. It requires more than learning to apply LEED checklists. Achieving true sustainability requires a new approach to creating and caring for the built environment.

GREEN BUILDING EXPERTISE

Green building requires new skills and new knowledge, as well as new attitudes and new mindsets. In a linear and hierarchical practice, each participant does his or her part and passes the job on to the next in line. There is little interaction, and people are compartmentalized by discipline or profession. By contrast, the green building process is interdisciplinary, iterative, and collaborative. Teamwork and critical thinking are valued. Everyone needs to learn to ask the right questions and to participate in developing the solutions. Feedback loops are built into the entire process.

The new skills required for a green building practice are not just knowledge of new strategies, materials, or equipment, although these are necessary. Green building practitioners need to learn how teams work, how to facilitate or participate in a productive discussion, how to work with people with different backgrounds and skills, and how to think outside their normal comfort zones when developing ideas. They need to be able to understand an ecologist's report on the proposed site, or better still, participate in walking the site and contributing to the assessment. They need to be able to question one another—Why should something be done the way it always has been done it in the past?—and then consider, what if…?

These are not skills and knowledge that most practitioners traditionally receive during their professional education and training. Most architects, engineers, landscape architects, planners, and business managers learn the new skills on the job and through trial and error, such as by facilitating meetings with team members and stakeholders. These opportunities will be explored in greater depth in Section 3. Additionally, training programs can help build these skills by combining experience with more formal classes, workshops, and online education. University curricula are beginning to incorporate these skills, but it may be several years before green expertise becomes the norm.

This guide is intended to set the foundation needed to develop green building expertise. The new understanding will change the way we look at the buildings we live and work in, the ones we walk past, and the ones we revere as beacons of innovation in our communities. It will challenge you to imagine the next green building project to which you'll contribute.

SECTION 2
SUSTAINABLE THINKING

Green building will change the way you think. Buildings that seem to be individual, static objects will reveal themselves as fluid systems that exist in relationship to their environments and change over time. Professionals who previously appeared only distantly related will become partners in a dynamic process that incoporates perspectives from different fields.

> **NO PROBLEM CAN BE SOLVED FROM THE SAME LEVEL OF CONSCIOUSNESS THAT CREATED IT.**
>
> Albert Einstein

This section reviews three major concepts that are integral to green building and sustainability: systems thinking, life cycle thinking, and integrated processes. In **systems thinking**, the built environment is understood as a series of relationships in which each part affects many other parts. Systems include materials, resources, energy, people, and information, as well as the complex interactions and flows between these elements across space and through time. Green building also requires taking a **life cycle approach**, looking at all stages of a project, product, or service. It requires asking, where do building materials and resources come from? Where will they go once their useful life ends? What effects do they have on the world along the way? Questions such as these encourage practitioners to ensure that buildings are adaptable and resilient and perform as expected while minimizing harmful consequences. Finally, to achieve results that are based on whole systems across their entire life cycle, building

professionals must adopt an **integrated process**. This approach emphasizes connections and communication among professionals and stakeholders throughout the life of a project. It breaks down disciplinary boundaries and rejects linear planning and design processes that can lead to inefficient solutions. Although the term integrated design is most often applied to new construction or renovations, an integrated process is applicable to any phase in the life cycle of a building.

In green building, solutions are examined through different perspectives, scales, and levels of detail, and then refined. The lens of each discipline involved in a project contributes to an overall view that leads to refined and more effective designs. For example, sustainable neighborhood design strategies might be analyzed by land-use planners, traffic engineers, civil engineers, infrastructure designers, public health experts, and developers. The more each team member understands the perspectives and strategies of the others, the more integrated the design. The iterative pattern of an integrated process can be used throughout the project as details come into focus. Far from being time consuming, the process can actually save time by encouraging communication up front and bringing people together for highly productive collaborative work sessions.

INTEGRATED DESIGN MEETS THE REAL WORLD

In the article "Integrated Design Meets the Real World," the authors note that users of an integrated approach "... got better at the process over time, especially when they were able to work with the same team members more than once, Once they'd gone through the process, they found it valuable, and many couldn't imagine doing design any other way."[15]

This section addresses problem-solving approaches that can be applied throughout the green building process. Subsequent sections will explore how green building professionals can begin to incorporate these ideas into projects and professional pursuits.

SYSTEMS THINKING

Sustainability involves designing and operating systems to survive and thrive over time. To understand sustainable systems, we must further understand what we mean by systems.

A **system** is an assemblage of elements or parts that interact in a series of relationships to form a complex whole that serves particular functions or purposes. The theory behind systems thinking has had a profound effect on many fields of study, such as computer science, business, psychology, and ecology. Donella Meadows, Jørgen Randers and Dennis Meadows,

15 A. Wendt and N. Malin, Integrative Design Meets the Real World, *Environmental Building News* 19(5) (2010), http://www.buildinggreen.com/articles/IssueTOC.cfm?Volume=19&Issue=5.

pioneers in the study of systems and sustainability, describe this discipline in their book *The Limits to Growth*.

A system can be physically small (an ant hill) or large (the entire universe), simple and self-contained (bacteria in a Petri dish) or complex and interacting with other systems (the global trading system or a forest ecosystem). Systems rarely exist in isolation; even the bacteria in the Petri dish are affected by the light and temperature of the laboratory. The boundaries of a system depend on what we are looking at, and most systems are actually systems within systems. For example, the human body is made up of many interlinking internal systems, such as the musculoskeletal system, which interact with external systems, such as the natural environment.

> Our training taught us to see the world as a set of unfolding behavior patterns, such as growth, decline, oscillation, overshoot. It has taught us to focus not so much on single pieces of a system, as on connections. We see the elements of demography, economy, and the environment as *one planetary system*, with innumerable interconnections. We see stocks and flows and feedbacks and interconnections, all of which influence the way the system will behave in the future and influence the actions we might take to change its behavior.[16]

Many systems in the modern world are designed as **open systems**, into which materials and resources are constantly brought in from the outside, used in some way, and then released outside the system in some form of waste. For example, in most urban American communities, water, food, energy, and materials are imported into the city from sources outside the municipal boundaries. In fact, many of our materials and resources are imported from around the world. After they have been used inside the city, they are released as waste in the form of sewage, solid waste, and pollution. In nature, there are no open systems; dead and decaying matter become food for something else, and everything goes somewhere. There is no "away." By slowing the passing of materials and resources through the system and linking elements to form new relationships and functions, we can begin to mimic nature and design **closed systems**, which are more sustainable.

When designing buildings and communities, we must understand both the individual elements of the system and their relationships to each other as a whole. One decision may have a ripple effect. For example, improvements in the **building envelope**, the boundary between the exterior and interior elements of a building, can change the requirements for the mechanical system. Using better insulation or more efficient windows might allow for a smaller heating system. At the same time, reducing air infiltration can raise concerns about the indoor air quality. Envelope design can also be used to increase daylight into the space, affecting lighting design, heating, and air-conditioning as well as improving the quality of the indoor space. But envelopes designed for increased daylighting without consideration of glare and heat gain can create uncomfortable and less productive spaces. Even the interior finishes and furnishings can change the effectiveness of natural daylighting and ventilation strategies.

---

16   Donella H. Meadows, Dennis L. Meadows, Jorgen Randers, and William W. Behrens III. (1972). *The Limits to Growth*. New York: Universe Books.

> **OPTIMIZING COMPONENTS IN ISOLATION TENDS TO PESSIMIZE THE WHOLE SYSTEM—AND HENCE THE BOTTOM LINE. YOU CAN ACTUALLY MAKE A SYSTEM LESS EFFICIENT, SIMPLY BY NOT PROPERLY LINKING UP THOSE COMPONENTS ... IF THEY'RE NOT DESIGNED TO WORK WITH ONE ANOTHER, THEY'LL TEND TO WORK AGAINST ONE ANOTHER.**
>
> Paul Hawken, Amory Lovins, and L. Hunter Lovins
> *Natural Capitalism*

The concept of feedback loops helps explain how systems work. **Feedback loops** are the information flows within a system that allow that system to organize itself. For example, when a thermostat indicates that the temperature in a room is too warm, it sends a signal to turn on the air-conditioning. When the room is sufficiently cooled, the thermostat sends a signal for the air-conditioning to stop.

This type of feedback loop is called a **negative feedback loop** because embedded in the system's response to a change is a signal for the system to stop changing when that response is no longer needed. Negative feedback loops enable a system to self-correct and stay within a particular range of function or performance. Thus, they keep systems stable.

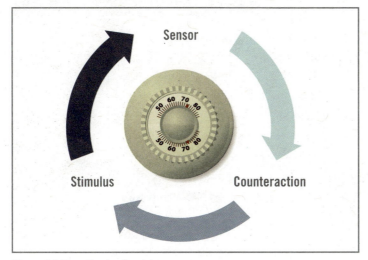

Figure 2.1. Negative feedback loop

Positive feedback loops, on the other hand, are self-reinforcing: the stimulus causes an effect, and the effect produces even more of that same effect. Population growth is a positive feedback loop. The more babies who are born, the more people there will be in the population to have more babies. Therefore, the population can be expected to rise until acted upon by another force, such as an epidemic or shortage of resources.

In the built environment, roads and infrastructure built out to the urban fringe

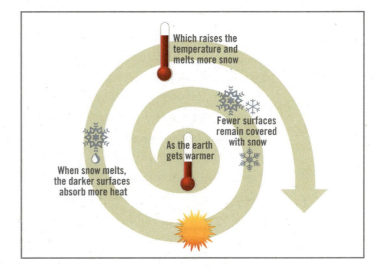

Figure 2.2. Positive feedback loop

often result in a positive feedback loop of increased development. This suburban growth can sprawl far from the urban core, requiring more roads and encouraging additional growth, as well as using more resources (energy, water, sewage systems, materials) to support that growth.

Climate change is another positive feedback loop. As the earth gets warmer, fewer surfaces remain covered with snow, a reflective surface that bounces incoming heat from the sun back into space. When snow melts, the darker surfaces absorb more heat, which raises the temperature and melts more snow. Similarly, in the built environment, the dark surfaces of roofs, roads, and parking lots absorb more heat from the sun. This **heat island effect** raises temperatures in urban areas several degrees above the temperature of surrounding areas, increasing the demand for cooling and the amount of energy that buildings use. The additional energy use can increase carbon emissions, which contribute to global warming, further raising urban temperatures and energy use, and the cycle continues.

Figure 2.3. Induced growth over time

Unchecked, positive feedback loops can create chaos in a system. For example, if urban temperatures rise too high, local populations may suffer or abandon the area. In nature, positive feedback loops are typically checked by stabilizing negative feedback loops, processes that shut down uncontrolled growth or other destabilizing forces. Stability and resilience in the system return as the feedback loops begin to control the change. To design sustainable systems, we must understand the positive and negative feedback loops already in existence or those we set in motion, to ensure systems remain stable and habitable over time.

Feedback loops—positive or negative—depend on flows of information. When information about the performance of the system is missing or blocked, the system cannot respond. For example, buildings without appropriate sensors and control systems cannot adjust to changing temperatures and maintain a comfortable indoor environment. The information must be both collected and directed. Most buildings have thermostats to provide information and control temperature. However, there are many other parameters, measurable or quantifiable characteristics of a system, that are relevant to sustainability but do not get measured or reported in effective ways. For example, the amount of energy used by tenant-occupied buildings may be collected by an electricity or gas meter and reported to the utility company but not to the occupants, who therefore have no information about their energy consumption and no incentive to reduce it. If real-time information on energy use is delivered to them in a convenient way, they can use energy more efficiently. Some have called

this the Prius effect, after the hybrid car that gives the driver information about fuel consumption so that she can drive in a fuel-efficient way.[17] Installing real-time energy meters where operators can act on the information is an example of connecting elements of a system so that they can interact and respond to each other more appropriately in the feedback loop.

In addition to elements, their relationships, and the feedback loops among them, systems theory explores the emergent properties of a system—patterns that emerge from the system as a whole and are more than the sum of the parts. For example, the pattern of waves crashing along the beach is an emergent property: the pattern is more than the water molecules that make up the ocean, more than the surface of the shore, more than the gravitational pull of the moon or the influence of the wind. The waves emerge as a result of the interactions and relationships among the elements.

Similarly, the culture of a company emerges from the people who work there, the buildings in which they work, the services or products they provide, the way they receive and process information, the management and power structure, and the financial structure. These elements and flows combine in both predictable and unpredictable ways to form a unique and individual organization. The elements of the system (people, buildings), the flows within the system (of materials, money, and information), the rules that govern those flows (management and structures), and the functions of the system (providing goods or services, generating a profit) determine whether the company is a good place to work and will be sustainable over time.

To influence the behavior of a system, it is important to find the **leverage points**—places where a small intervention can yield large changes. Providing building occupants with real-time energy information is an example of using a leverage point to alter behavior. Rather than changing the elements of the system—the envelope of the structure, the mechanical system, the building occupants, the electricity grid—the change focuses merely on delivering available data to a point where it can be acted on appropriately. This minor tweak can dramatically raise the efficiency of the system. Donella Meadows's essay "Leverage Points: Places to Intervene in a System" provides an excellent summary of how to find and use leverage points to make meaningful change.[18]

In *Natural Capitalism*, Hawkens, Lovins, and Lovins explore how capital markets can be used for —rather than against—sustainability, not by eliminating them or adding intensive regulation, but by using leverage points within the system. One leverage point they examine is the goals that govern the system. By valuing not only financial capital but also natural capital and human capital, existing systems and structures can lead to sustainability.

17   Brand Neutral, *The Prius Effect: Learning from Toyota* (2007), http://www.brandneutral.com/documents/Prius_Effect.pdf.
18   D. Meadows, *Leverage Points: Places to Intervene in a System* (1999), http://www.sustainer.org/pubs/Leverage_Points.pdf.

Project Case Study

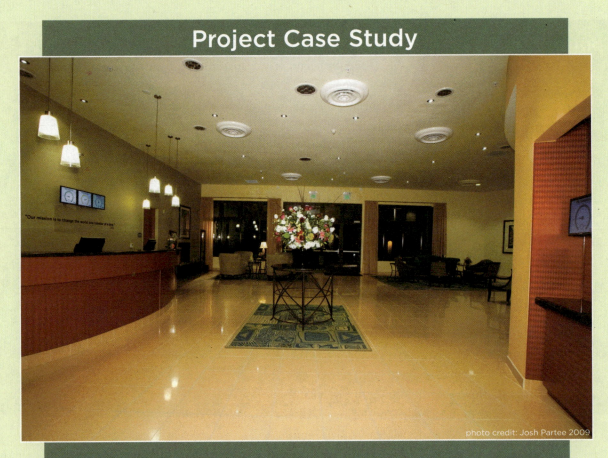

photo credit: Josh Partee 2009

GAIA NAPA VALLEY HOTEL

The Gaia Napa Valley Hotel in Canyon Valley, California encourages its employees and visitors to apply systems thinking. The hotel provides an interactive computer screen in its lobby that displays real time information about the building's water and energy use, as well as its carbon emissions. The interface makes the project's commitment to energy efficiency and developing a beautiful, functional and sustainable facility tangible and encourages visitors and employees to reduce their impact while at the hotel. Additionally, this display inspires visitors to reflect on their habits and consider making changes to their resource consumption once they return home. This type of interactive display helps educate occupants about the impact of green building, and support the Gaia Napa Valley Hotel's efforts to achieve Gold certification under LEED for New Construction, version 2.1. For more information about this project, please visit http://www. gaianapavalleyhotel.com/.

For instance, when carpet manufacturer Interface Flooring switched from being a producer of carpet to a provider of the service of floor coverings, it created a shift in the company's mission. Instead of buying carpet, customers could buy the *service* of the carpet, which would be owned by Interface. The company would be responsible for maintaining the carpet over time, replacing worn areas, and disposing of any "waste." This shift served as a leverage point to enable the company system to change radically toward sustainability, reducing waste, and improving performance of the product while maintaining profit. In other words, Interface Flooring moved from an open system to a closed system. The new mental model resulted not just in more efficient processes, but also in a radical restructuring of the company and all its operations.

Buildings are part of a world of nested systems that affect and are affected by one another. Once the project team understands the network of systems that affect a given project, the energy and matter that flow through the systems, and the relationships and interdependencies that exist, the deeper and more effectively integration can occur.

When designing aspects of the built environment, consider the systems in which the project will be

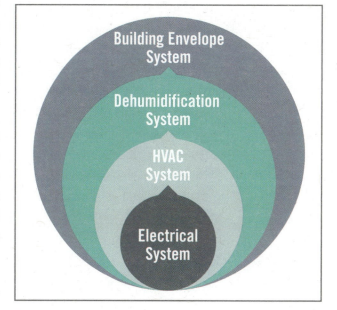

Figure 2.4. Nested systems

located and the systems the project will create. Learn about the relationships between the elements, the flows of resources and information, and the leverage points that can lead to dramatic changes. Before starting any project, the team can explore these systems by asking questions. Whether working in the planning, design, construction, or operations phase, these questions may provide insight into the systems context and ways to move more fully toward sustainability in an integrated way.

QUESTIONS A PROJECT TEAM NEEDS TO EXPLORE AS MEMBERS BEGIN WORKING TOGETHER, INCLUDE:

- Where is the project located, and who are its neighbors—locally, regionally, and beyond? What is the local watershed? The bioregion? What are the characteristics of these systems?

- How do resources, such as energy, water, and materials, flow into the project? Where do they come from, and from how far away? What other purposes or projects do those flows serve?

- What natural processes are at work on the site? How do resources, such as rainwater, wastewater, and solid waste, flow out of the system? Where do they go? Are there places on site where these flows can be captured, stored, or reused?

- What are the goals of the owner? What is the function or purpose of the project? How will the project meet those goals?

- What is the community within the project? Who are the people who come here, and where do they come from? Where do they go? What brings them together, and what might keep them apart? How will the project change their interactions?

- How does the project community interact with other, overlapping communities? What are the interrelationships? Are there sources of conflicts? What is the economic system within the project? How does it fit into larger or overlapping economic systems?

- What are the leverage points within the system? Are there places where small changes can produce big results?

In a linear design process, the solutions to one problem may cause other problems elsewhere in the system. When problems are solved through a systems-based approach, multiple problems can often be solved at the same time. This synergy is possible when we take the time to explore the interconnections and approach a project in a holistic manner. In the context of the built environment, systems thinking allows us to explore and support the rich interactions that make healthy, thriving, and sustainable communities.

LIFE CYCLE APPROACH

Green building takes a **life cycle approach**, looking at the entire life of a project, product, or service, rather than a single snapshot of a system. The dimension of longevity distinguishes green building from conventional building practice, which may fail to think across time, and helps create communities and buildings that are meant to last. For a building, a life

cycle approach begins with the initial predesign decisions that set goals and a program to follow. It continues through location selection, then design, construction, operations and maintenance, refurbishment, and renovation. A building's life cycle ends in demolition or, preferably, reuse.

In most cases in our industrial system, we treat the manufacture of products, the construction of buildings, and the operations of organizations as open systems. We take materials from outside the system, use them to make something, and then discard what remains. This throughput of resources occurs at every phase of the life cycle, creating a constant cycle of consumption and waste. In addition to the upstream effects that happen before a material is used, there are downstream impacts associated with its operation and end of life. We need to consider both upstream and downstream effects in our decision-making processes.

Systems thinking relies on identifying and acting on opportunities to close this loop. Because we typically do not consider building elements as linked into a larger set of systems, this waste remains largely invisible. By incorporating the upstream effects into our analysis of alternatives, we can get a broader picture of the environmental costs and benefits of materials. The practice of investigating materials from the point of extraction to their disposal is sometimes described as **cradle to grave**—a term that suggests a linear process through an open system. To emphasize the cyclical aspect of a closed system, architect William McDonough and colleague Michael Braungart coined the phrase **cradle to cradle**. In a closed system, there is no waste, and all things find another purpose at the end of their useful lives.

A comprehensive, life cycle approach improves the ability to address potentially important environmental and human health concerns. For example, a product may consist of material mined in Africa, manufactured in Asia, and shipped to the United States for purchase. By focusing only on the energy efficiency of this product during its use, we might miss the damage caused by its transport from the place of manufacture or by the extraction of its raw material. Or a window may have a high recycled content but not be highly efficient.

By looking only at the percentage of recycled content, we might select a product that will compromise the project's energy-saving goals. In a green building project, the team must consider **embodied energy**—the total amount of energy used to harvest or extract, manufacture, transport, install, and use a product across its life cycle— alongside performance and adaptability. The careful consideration of all attributes may lead to the selection of products that did not at first appear to be the most sustainable option.

Life cycle thinking can be applied to environmental considerations, in which case it is called **life cycle assessment (LCA)**, and to cost considerations, or

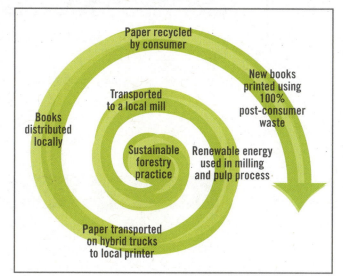

Figure 2.5. Considering a product's entire life cycle.

Within the figure:
Paper recycled by consumer
Transported to a local mill
New books printed using 100% post-consumer waste
Books distributed locally
Sustainable forestry practice
Renewable energy used in milling and pulp process
Paper transported on hybrid trucks to local printer

life cycle costing (LCC). These are distinct approaches with different methodologies but are often confused. Both can support more sustainable decision making, but they use different types of data and provide different kinds of information.

Life cycle assessment attempts to identify and quantify environmental effects throughout the life of materials, products, or buildings. It identifies all the processes and associated inputs (energy, water, materials) and outputs (wastes, by-products), from the extraction and processing of raw materials and recycled feedstocks, the transportation of these materials, and the manufacturing and packaging of the product to its use, maintenance, and finally its recycling or disposal. These inputs and outputs are quantified and their effects on the environment and human health are measured. Although LCA does not address all potential effects, it provides a comprehensive picture of the life cycle. This information can then be used to support decision making. Tools and databases used in conducting LCAs are available from sources in the U.S. government and the private sector.

Life cycle costing looks at both purchase and operating costs as well as relative savings over the life of the building or product. It calculates payback periods for first costs, providing a context for making decisions about initial investments. For example, more efficient mechanical systems generally cost more than inefficient equipment, but by looking beyond the purchase price and calculating all the energy, maintenance, replacement, and other costs over the life cycle of the equipment, we can better understand the true cost of the equipment—both to the environment and to the building owner.

LCC can be used in comparing alternatives with different initial and operating costs. For a building this usually includes the following costs:

- Initial purchase, acquisition, or construction
- Fuel
- Operation, maintenance, and repair
- Replacement
- Disposal (or residual value for resale or salvage)
- Finance charges
- Other intangible benefits or costs, such as increased employee productivity

Life cycle thinking can be applied to all decisions in green building, not just products and buildings. Teams need to look for opportunities to evaluate the environmental impacts of design decisions and improve sustainability at all points in the project's life cycle. Once decisions have been made at each phase, however, those opportunities can become limited. The key to sustainability is to establish goals and targets early in the process, understand the systems that are in play, and anticipate how those systems are likely to change and evolve.

Land-use and urban planners also draw on the concept of life cycles because decisions about the location of roads and infrastructure can affect all future decisions about that land for centuries. Consider again the example from Section 1 of Rome's road structure: these roads were built for pedestrians and therefore remain walkable and pedestrian oriented even today. This does not mean that there are no opportunities to make vehicle-oriented development

greener, but it does mean that the challenges of reducing transportation impacts, such as carbon footprint, are greater in projects where pedestrian access is not an initial goal.

With future implications of the built environment in mind, we must rethink the processes we use at all phases of the life cycle. Assembling the right team, establishing goals, and understanding the systems and metrics for success will help ensure that we move closer to a sustainable built environment.

INTEGRATED PROCESS

Integrated design is the current buzz word in the green building world—even though few can say precisely what it means. An **integrated process**, as it relates to green building, is an interdisciplinary method for the design and operation of sustainable built environments. The integrated process builds on the two previous principles addressed in this section, systems thinking and a life cycle approach. Although practitioners often refer to integrated *design*, the integrated *process* can be used for all stages of a green building project, from design and construction to operations and reuse or deconstruction.

An integrated process provides opportunities to consider resources in new ways. It encourages professionals to think and make decisions holistically. For example, in the conventional building design process, hydrologists, civil engineers, mechanical engineers, and landscape designers all make decisions involving water. Often, though, these professionals make their plans for potable water use, irrigation needs, wastewater disposal, and stormwater management separately. In contrast, an integrated process is highly collaborative. Conventional planning, design, building, and operations processes often fail to recognize that buildings are part of larger, complex systems. As a result, solving for one problem may create other problems elsewhere in the system. For example:

- Separating residential and commercial uses and failing to connect them with alternative transportation means that people will drive cars to reach their destinations, generating air pollution and traffic
- Filling a landscape with ornamental plants not appropriate for the local climate means that large amounts of water may be required throughout the life of the project
- Creating air-tight buildings for energy efficiency without providing adequate ventilation results in poor indoor air quality for building occupants

When an integrated, systems-based approach is used, the solution to one problem can lead to solutions to many problems. The process of planning a project's water use might lead to the design of systems that capture rainwater and greywater to meet water supply and irrigation needs while reducing runoff and protecting water quality. More broadly, by thinking about the system across the entire life cycle, integrated strategies can be developed synergistically.

For example:

- Locating homes near jobs and shops and designing safe, pedestrian-friendly streets can encourage people to walk, both reducing vehicle emissions and improving their health
- Designing landscapes that use native species can both reduce water consumption and provide habitat for local fauna
- Orienting buildings appropriately on a site and designing them to catch sunlight for heating and illumination and natural breezes for cooling and ventilation can save energy, improve indoor air quality, and even increase workers' productivity
- Composting improves the quality of the soil and reduces greenhouse gas emissions related to trash hauling

Practitioners of an integrated process must develop new skills that might not have been required in their past professional work: critical thinking and questioning, collaboration, teamwork and communication, and a deep understanding of natural processes. An integrated process is a different way of thinking and working, and it creates a team from professionals who have traditionally worked as separate entities.

The integrated process requires more time and collaboration during the early conceptual and design phases than conventional practices. Time must be spent building the team, setting goals, and doing analysis before any decisions are made or implemented. This upfront investment of time, however, reduces the time it takes to produce construction documents. Because the goals have been thoroughly explored and woven throughout the process, projects can be executed more thoughtfully, take advantage of building system synergies, and better meet the needs of their occupants or communities, and ultimately save money, too. The specific steps involved in the integrated process will be addressed in Section 3.

Nature has much to teach us about applying systems thinking, a life cycle approach, and integrated processes to our work. By observing natural patterns, such as how heat flows, water moves, or trees grow, we can learn to design systems that use resources effectively. The fields of biomimicry and permaculture provide two different and innovative approaches to solving problems by following nature's patterns and strategies. Both of these fields of practice ask: how would nature solve this? Similarly, green building practitioners can use the core concepts addressed in this section to determine the nature of the systems in which they are working, meet the needs of the community, and set goals and priorities for the project.

SECTION 3

SUSTAINABLE THINKING AT WORK: NEW PROCESSES FOR BUILDING GREEN

Green building requires a new way of thinking and approaching the design, construction, operation, and renovation of buildings and communities. Basic elements of this approach were presented in Section 2. The concepts of green building are valid for many types of buildings at all stages of development and questions will likely arise as you begin to apply them. How do teams organize as part of an integrated process? How does systems thinking change the way sites are developed? How does life cycle assessment affect materials selection? In short, how does this new approach work in real life?

This chapter focuses on the processes surrounding green building—*how* these concepts can change the way you do things—and describes successful approaches to green building, with case examples of actual projects. The strategies and technologies of green building—*what* is done—will be discussed in Section 4.

GETTING STARTED

Several principles form the foundation for successful practice:

Process matters. How you approach projects is crucial to what you do and are able to accomplish. In other words, a good process is essential to good outcomes.

Get in early. The commitment to green building should be made as early as possible so that it can assist in framing effective goals. Trying to add green features to a project late in the process is the most expensive and least effective approach. For community or neighborhood projects, the commitment should be made at the beginning of the land-use planning phase so that it can inform land-use decisions and zoning, design of transportation systems, and layout of infrastructure. For new construction, early means before the site is selected and before the team is selected, if possible. For operations and maintenance projects, commitments need to be established before any action toward change is taken.

Follow through. The commitment to green needs to continue throughout the life of the project. The green building process does not end when the project team hands the site over to the owner, facility manager, or tenant. Follow-through is needed at all stages to ensure that the strategies and technologies are maintained or adapted as necessary to remain effective. Additionally, ongoing training ensures knowledgeable operation and maintenance of these strategies and technologies, as well as an opportunity to provide feedback on the challenges faced and lessons learned.

Look beyond first costs to long-term savings. This new process doesn't typically cost more, but it does shift costs earlier. Increased efficiency and savings come later. Up-front goal setting, analysis, and evaluation of alternatives will assist in making decisions that result in savings over the long term through synergies and integration. Synergies are actions that complement each other, creating a whole greater than the sum of its parts. The savings are often reflected in life cycle costing. Green strategies and technologies often have very short payback periods, but when organizations budget planning and design costs separately from capital projects and operations, savings in one category may not provide a persuasive argument for increased spending in another. It might be necessary to bring the stakeholders from these departments together to establish mechanisms for interdepartmental and collaborative decision making and funding.

Include and collaborate. Green building demands that a multidisciplinary team of professionals join with members of the community involved or affected by the project to look at the big picture, not just the individual elements that concern each of them most immediately.

ESTABLISHING AN ITERATIVE PROCESS

All the activities described in this section take place in an iterative process that contains numerous feedback loops. An **iterative process** is circular and repetitive. It provides opportunities for setting goals and checking each idea against those goals.

An iterative process has a cyclical nature:

- Establish clear goals and overarching commitments
- Brainstorm and develop creative solutions
- Research and refine ideas

- Explore synergies between specific strategies
- Establish metrics for measuring success
- Set new goals based on the work that has been done

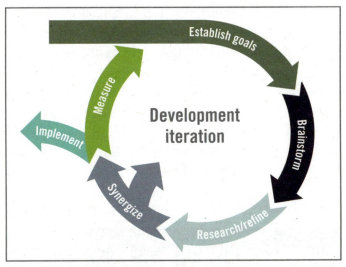

Figure 3.1. Iterative Process

This is a way for project teams to apply systems thinking and integrated process. It differs from traditional processes in that it is not linear, as when one team member completes a task and passes the work off to the next person. Instead, the team works together, in small groups and as a whole, to develop the project design and plan collaboratively. Ideas are continually being developed by the entire team, researched and refined by smaller groups, and then brought back to the team to consider critical next steps and make final decisions.

In early project meetings, it is important to establish a common commitment to the planning and implementation process. Some team members might not be familiar with an iterative process. Even when the team is experienced, it is worth reviewing the steps to ensure that all team members understand it in the same way—perhaps by asking how they might approach a problem. Sometimes the iterative process involves looking deeply at why or how a specific idea would work; at other times the team will compare one strategy with others to explore synergies and trade-offs.

Defining critical milestones, assigning champions, and clarifying goals up front will enable projects of all sizes and types to incorporate sustainability more effectively. Over the course of a project, especially a long and complex one, goals and targets evolve. Through the iterative process, a team can be ready to address changes and make deliberate decisions by using information from smaller group meetings.

An experienced facilitator can encourage people to voice their thoughts. A facilitator assists the team in expressing new ideas and ensuring that varying perspectives are valued. Additionally, this person brings the group back to explore how proposals will either further or hinder achievement of the project goals.

Careful documentation helps capture the lessons learned on the project so that they can be applied in the future—either within the timeline of that project or on subsequent green building projects.

Many different types of meetings may be useful in an iterative process. Although approaches will vary based on the specific project and team, the process often includes charrettes, team meetings, small task groups, and stakeholder meetings.

Charrettes are an important tool in an iterative process. Named after the carts that carried French architecture students' models to their final review (often as the students frantically completed their work en route with the help of friends), charrettes are intense workshops designed to produce specific deliverables. A charrette brings together the project team with stakeholders and outside experts as needed for creative thinking and collaboration. Generally held at the beginning of the project, charrettes assist in establishing goals. These sessions can also be held throughout the project at major milestones for focused, integrated problem solving. They energize the group and promote trust through productive dialogue. Additionally, they ensure alignment around goals, objectives, and actions. Although we typically think of "design charrettes," charrettes can be used for all types of building projects.

STAKEHOLDERS

The term stakeholder encompasses more than just decision makers and includes those who must live with the decisions and those who must carry them out. This cross-section of perspectives depends on the type of project. Participants in a design-build project might include the building owner, developer, client, design team members, facility managers, community representatives, local regulatory agencies, local environmental groups, ecologists, and tenants or other building users. Operations projects might also include cleaning contractors, waste management contractors, landscape contractors, local real estate and leasing specialists, and salvage and resale companies.

Charrettes derive their value from the collaboration of people from different disciplines and perspectives. When setting up charrettes, then, include all relevant stakeholders and experts. Those outside the project team, particularly stakeholders in the community, might need encouragement to attend and a commitment that their voices will be heard. One-on-one conversations prior to the event are often useful in gaining initial trust and confidence. An educational component can ensure that participants with varying levels of knowledge all have an adequate understanding of the topics under consideration.

The combination of brainstorming, different perspectives, and a focus on results distinguish the charrette from other types of meetings. Because charrettes are highly structured, they require a strong facilitator, who may come from outside the core project team. The ideal candidate is an excellent listener who can distill the big picture from multiple viewpoints. It is critical that this person guide the conversation in a productive and unbiased direction.

Since charrettes are generally designed to result in a concrete product, an agenda and clear goals are needed. Discussion questions and activities must be designed to meet those goals. However, the charrette also needs to be flexible enough to allow for the emergence of extraordinary ideas. In advance, the project owner or developer may draft a statement that establishes the goals of the charrette and its relevance to the project. The statement inspires the team to reach the goals and also assures participants that their work is important and will influence the final project. Clear goals and specific deliverables and outcomes help all participants understand the purpose of the charrette and set the foundation for an effective agenda. Each agenda needs to be tailored to the specific project, but in general, a charrette takes the following form:

- Background briefing, to ensure that all participants have the basic information on the project and topics to be discussed
- Brainstorming, small-group work, reports, further brainstorming, and subsequent reports structured around discussion questions and specific tasks
- Synthesis of work, development of recommendations, and identification of deliverables
- Initial response from the owner or developer to the recommendations, affirming the commitment to sustainable approaches and ideas
- In follow up, a written report documenting the charrette and identified action items should be sent to all participants

Team meetings can allow the group to work together creatively on new synergies. For example, the development of an integrated water conservation system might require collaboration between the landscape architect, the civil engineer, the structural engineer, and the mechanical, electrical, and plumbing (MEP) designer. Meetings are more effective if facilitated by a neutral party who encourages all team members to speak up.

Small task groups provide opportunities to explore particular topics, conduct research, and refine the ideas for presentation at a later team meeting. They are generally composed of existing team members but may require outside experts. They do not need to be multidisciplinary unless appropriate for the task. Task group members should view their work as exploratory and consider all ideas, even those that appear to be poor choices or infeasible. Investigation of high-risk ideas can lead to the most innovative aspects of a project. Many of the specific strategies discussed in Section 4 of this guide require task groups to flesh out ideas and determine appropriate application.

Stakeholder meetings are held with neighbors, community members, and others with a vested interest in the project. They enhance a project team's interaction with and understanding of community issues, concerns, and ideas. Local residents frequently bring a deep understanding of the place—the local context, culture, and history, as well as the strengths and needs of the community.

In most communities, it is essential to win the trust of local residents and organizations, which may involve one-on-one and small-group meetings. It is easy for a project team to underestimate the value of this step and instead call an evening meeting with the community to present the proposed project. Effective stakeholder meetings involve both careful listening and openness to determine the most feasible and effective solutions for the community.

As with any break with tradition, barriers and obstacles can arise when a team uses an iterative process. In the article "Integrative Design Meets the Real World," authors Wendt and Malin highlight the benefits of the integrative design process but also discuss some of the obstacles:

- Meetings can be expensive to run and hard to schedule
- Communication between meetings often breaks down
- People may be resistant to green goals

- Participants can balk at the iterative, integrative process
- Traditionalists may resist the up-front loading of modeling, testing of assumptions, and analysis
- People may be reluctant to embrace new technologies[19]

Importantly, experts interviewed for the article noted that they got better at the process over time, especially when they were able to work with the same project team members on more than one project.

TEAM SELECTION

THE MASTER BUILDER

Master builders were schooled through local apprenticeships, and the techniques and technologies they learned were developed from an understanding of local issues and passed down through generations. Mechanized transportation was limited, so people possessed an intimate knowledge of local materials as well as workforce skills, economics, cultural imagery and traditions, microclimates, and soil conditions. They understood the flow of local resources and what local conditions could be limiting. The built environment was designed and constructed from a deep connection to each individual place, with the master builder conceptualizing the overall pattern and each artisan, craftsman, and journeyman then contributing layers of richness and diversity at smaller scales. What resulted were buildings and communities that truly were integrated with their environment and that lived, breathed, and grew to become timeless elements of their place.[20]

One defining element of the green building process is the **project team**, a broad, inclusive, collaborative group that works together to design and complete the project. This team differs from the group of stakeholders who participate in the charrettes. The members of this group are highly invested and involved across all stages of the project. They are deeply involved in the problem-solving and decision-making processes at every step.

Individual projects require different blends of expertise. For example, the appropriate team for developing a sustainable operations program would likely involve the facility owner, facility management team, vendors, occupants' representatives, and a sustainability expert. Additionally, the expertise of individual project team members will be more critical at different points in the project. For example, an ecologist might be most relevant during the initial stages of the project, to help the team understand and work with the site, but could bring forward valuable ideas and find synergies throughout the process.

19   A. Wendt and N. Malin, Integrative Design Meets the Real World, *Environmental Building News* 19(5) (2010), http://www.buildinggreen.com/articles/IssueTOC.cfm?Volume=19&Issue=5.
20   7group & Reed, B. (2009). *The Integrative Design Guide to Green Building: Redefining the Practice of Sustainability.* Hoboken, NJ: John Wiley & Sons, Inc.

Project Case Study

photo credit : Josh Partee 2009

KENYON HOUSE

As a two-story, 18-unit residence providing affordable housing for formerly homeless people living with HIV and AIDS, the Kenyon House project in Seattle, Washington faced many unique challenges in its effort to build green. Surprisingly, though, one of the greatest barriers the project team faced in developing a LEED-certified facility was the city zoning requirement that allowed for no more than 50 feet of street frontage. This code would have prevented the project from achieving many of its sustainability goals through orienting for maximum daylighting and solar gain. John Woodworth, Principal at SMR Architects, attended neighborhood meetings and worked to inform stakeholders on the benefits of the sustainable development. By including stakeholders in the process, the group filed a well supported zoning complaint. The project was able to move forward with development, unhindered by zoning issues and with full support of those that became involved in the development process. This sort of collaborative approach to solving problems supported the Kenyon House in earning LEED Platinum certification under the LEED for Homes rating system, version 1.0. You can learn more about Kenyon House at www.usgbc.org/casestudies.

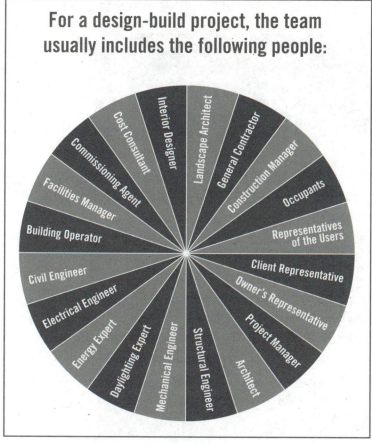

For a design-build project, the team usually includes the following people:

Labels around the wheel: Interior Designer, Landscape Architect, General Contractor, Construction Manager, Occupants, Representatives of the Users, Client Representative, Owner's Representative, Project Manager, Architect, Structural Engineer, Mechanical Engineer, Daylighting Expert, Energy Expert, Electrical Engineer, Civil Engineer, Building Operator, Facilities Manager, Commissioning Agent, Cost Consultant

FIGURE 3.8. Members of an Integrated Team

The team process favors a design-build or integrated project delivery (IPD) contracting process rather than traditional design-bid-build, in which the contractors are brought in after many elements of the project have been determined. Design-build and IPD enable team members to participate from the early project stages, including goal setting and initial brainstorming.

Team members should understand green building and have experience participating in a team. The experience and commitment to sustainability needs to extend to subcontractors and trades as well. Requests for proposals and interviews should include questions about experience in green building and sustainability. Ideally, evaluation of bids is based on the *best low* bid rather than the *lowest* bid. Even when this is not possible, as on many public projects, prerequisites identified in the RFP can help ensure that teams are qualified. Specific qualifications to look for might be past participation in integrated design processes, experience on green or LEED-certified projects, and LEED professional credentialing, from LEED Green Associate to LEED Accredited Professional. If inexperienced people are on the team, some training and orientation to the process will be necessary.

GOAL SETTING

This guide repeatedly emphasizes the importance of project goals; every green building project needs to be grounded in strong goals and set a clear pathway to ensure they are achieved. Clear goals articulate what the project will be designed to accomplish, by:

- Making sure that the vision is clear
- Providing a frame of reference for the whole project
- Defining the sustainability targets and keeping the project on track to meet them

Setting lofty-sounding general goals can be tempting; however, such goals may not provide enough information to guide a project. For example, saying that a project should be "healthful" may be appealing, but what does that really mean in the project context? How will you know if you are on the right track? This type of high-level goal needs to be accompanied by *metrics*, things that can be measured, and *targets*, levels of achievement that should be reached. Each

goal may have multiple metrics and targets. For example, if by "healthful" the team means that the project should protect indoor air quality, one metric for that might be the amount of volatile organic compounds in building materials. A target associated with that metric might be that all paints have zero VOCs. There are many attributes to indoor air quality, so in addition to addressing the potential sources of pollutants (such as materials that emit VOCs), the team must develop metrics and targets for proper ventilation.

Project goals and their associated metrics and targets can be both quantitative and qualitative. For example, if a goal is that a neighborhood project be walkable, a team might consider as a quantitative measure the percentage of homes that are within a quarter-mile of destinations such as parks, restaurants, and stores. They also might consider qualitative factors, such as whether the project has functional sidewalks. This metric is qualitative because the presence of sidewalks doesn't necessarily contribute to walkability. The usage of those sidewalks, however, can demonstrate the walkability of the neighborhood. Another example: the goal of a waste management program in an existing building might be to make recycling convenient. The quantitative metric might be the number and location of recycling receptacles and the ratio of receptacles to employees on site. A qualitative factor might be the usage of recycling receptacles: are those adjacent to workspaces and offices used more than those at central locations, such as break rooms, or vice versa? Such assessments can help the team achieve its goal through changing the placement or number of receptacles.

In addition to being measurable and accompanied by appropriate metrics and targets, effective *goals must be achievable*. Goals that are completely out of reach because of cost or available technology do not provide guidance and can lead to frustration. On the other hand, goals that articulate aspirations will provide a challenge that inspires the team to new heights. For example, "to stop global climate change" is an unachievable, ineffective goal. Similarly, if the project is in an existing building with limited roof area and a limited budget, "to achieve net-zero energy" is unrealistic because the building cannot accommodate on-site energy generation or be redesigned with no mechanical system. In both cases, a better project goal might be "to avoid contributing to greenhouse gas emissions." The team could achieve this goal by reducing the project's energy use and offsetting emissions by purchasing renewable energy credits.

Goals should reflect the *spatial scales and time horizons* that the project can affect, assuming a realistic rate of change. Stopping global climate change is beyond the space and time constraints of a single project. Even "to reduce greenhouse gas emissions by 30%" may be impossible for a project to implement all at once. Therefore, many climate-related targets are written, "to reduce greenhouse gas emissions by 30% by 2030." This type of time horizon is particularly appropriate for very large or complex projects, such as cities, organizations with multiple locations, and large campuses, where there are many different sources of greenhouse gases and time is needed to develop and implement sufficient reduction measures and policies. For example, in 2010, the Obama administration announced a target for the U.S. federal government to reduce its emissions by 28% by 2020.

Systems thinking and integrative principles encourage setting goals that go beyond deciding to seek specific LEED credits or a specific certification level. Although some teams use green building checklists, such as the LEED checklist, as the basis for setting project goals, projects are likely to be most successful if goals reflect why the project is being undertaken and how success will be demonstrated and measured. Once these goals are articulated, checklists can serve as the basis for making decisions throughout the process.

Since it is crucial to reach an agreement on the project goals, a charrette, perhaps followed by a series of team meetings, is recommended. The number of meetings will depend on how complex the project is and how quickly alignment can be reached by the stakeholders. Before these meetings, the project owner should think about underlying goals for the project, why it is needed or wanted, and what it should achieve, and discuss these points with the facilitator. Next, the project team and major stakeholders should engage in an initial goal-setting discussion, building upon the owner's initial ideas. This session should include representatives of the community and other experts to provide information on local environmental, social, and economic issues.

Once the goals have been established, they need to be listed and described in a written report. Identifying a project team member as the "goal keeper" ensures that all subsequent work can be related to the goals. Different goals may require different champions, depending on the complexity of the project. For example, the role of the commissioning agent is to ensure that goals are articulated by the owner, understood by the design team, incorporated into the design, and then achieved during construction. Thus the commissioning agent is well positioned to follow the progress of the project in relation to established goals. Not every project has a commissioning agent, but that role can be played by other members of the team.

OBSERVATION OF THE SYSTEM

Getting to know the site is part of the needs assessment and evaluation process. This will help during the team's big-picture discussions of how to turn the goals into a concrete action plan. Design-build projects that can choose a site will benefit from setting goals before selecting a location for the project, thereby ensuring that the location contributes to the overall project plan rather than presenting challenges that the team must overcome.

The most obvious way to learn about a place is to spend time there, preferably at different times of day and in different seasons. By observing the place, people, wildlife, plants, and weather, team members can understand the patterns that make the place unique. Before they can do that effectively, from a sustainability perspective, they need to understand what is. This applies to existing buildings as well. The building ownership and management structure, use and users, and relationship to the community need to be taken into account. For instance, if the building has 32 tenants, installation of submeters in all data centers will have different implications than if it were a single-tenant facility. By studying the site, the team, with help

from the facilitator, can ensure the project's connection to the neighborhood.

Meaningful data gathering and interpretation often require the expertise of technical specialists, such as hydrologists, ecologists, engineers, economists, and anthropologists. There are many tools that can support this effort, such as systematic data collection and analysis and mapping. For existing buildings, information may be obtained through occupant surveys, building walkthroughs, and audits.

Geographical information systems (GIS) can help illustrate how different elements intersect and overlap. Map layers might show soils, infrastructure, shade, wind patterns, species distribution, land uses, demographics, roads and transit routes, traffic patterns, walkways and barriers, material flows, and solid waste pathways. Maps can also display growth projections, targeted development areas, and other indicators of how the site is likely to change over time.

It is important also to understand the patterns at work at different spatial scales. Mapping should always extend beyond the project borders to show how it fits into a local as well as regional scale. For example, the level of detail at a small scale might reveal much about the local street grid, but zooming out reveals connections to the regional transporation system.

Once all the relevant information about the project has been collected and assessed, it is time to return to the project goals. Given what the team has learned about the project systems, its needs and resources, do the goals of the project make sense? Are they achievable? Are there other ways to meet those goals by finding other leverage points in the systems? For a renovation project, the team might prepare a gap analysis that compares existing conditions with goals and identifies the gaps. Depending on what has been learned through observation, it may be necessary to go back and refine or revise the goals.

OBSERVING A SYSTEM

To observe and understand the site, team members must ask many questions:

- What are the general climatic patterns of the site? What are the microclimates, and how and why do they occur? How does water fall on and run off the site? How does the sun affect these conditions?

- What are the soils like on the site? Are they rich loam or hard clay? Has the site ever been used for agriculture? Can it be used to grow food now?

- What plants and animals exist on the site? How did they get there? Are they healthy or stressed?

- How does energy get to the site? Is the site remote or connected to a utility grid?

- Are there roads? What type? Where do they go? Do they have sidewalks? How do the current occupants use this infrastructure?

- What kind of buildings are on the site? How tall are they? How do they connect to the street? Are they new or old? Occupied or vacant? What are they used for?

Project Case Study

photo credit: Josh Partee 2009

CHARTWELL SCHOOL

Those involved in the early development of the Chartwell School in Seaside, California knew that their ultimate goal was to create an environment that would dovetail its sustainability efforts with its educational objectives. Having established this goal, a program document was created to explain how sustainable building and a positive learning environment could be conjoined. Knowing that natural light creates a positive atmosphere for learning and also decreases a building's needs for electricity and lighting resources, the project team found innovative ways to accomplish ambitious goals such as daylighting every space in the building. This helped them attain other project goals, including achieving LEED Platinum certification and developing a net zero building. For more information about the Chartwell School, please visit their website at http://www.chartwell.org/.

EXPLORATION AND SELECTION OF TECHNOLOGIES AND STRATEGIES

Sustainable design requires thinking methodically through the types of strategies for each aspect of the system and evaluating alternatives against project goals through an iterative process. Although this process may be more involved and more expensive than a conventional design process, it is more likely to help the team arrive at solutions that will serve the project owner, the occupants, and the community over time. In general, the evaluation and selection phase of a sustainable design process involves listing all types of strategies and technologies that might make sense. This broad list is then reviewed and options narrowed based on certain criteria, such as whether a strategy is feasible on the site, whether a technology is available, and whether an approach is appropriate for the project. Once the list has been narrowed, more focused analysis may be required.

For some projects, it may seem easy to list the alternatives and then decide on the best one. For example, when designing a new waste management program in a town that has only two waste haulers, the choice may seem simple. But even this situation requires a thorough investigation. The team would first collect all the relevant information about the two waste haulers. They might find that one costs less but that the other has a higher recycling **diversion rate**,[21] the percentage of waste materials diverted from traditional disposal methods and recycled, composted, or reused. One hauler may accept only sorted recyclables, but the team has determined that a commingled program is more appropriate for the project occupants. Choosing between these two based on this information would require revisiting the team goals. But what if the team values both recycling and cost savings? Or what if another goal is to reduce the greenhouse gas emissions associated with solid waste? The team would then have to consider additional information, such as the distance of each waste management facility from the project site, the types and sizes of trucks used for hauling, and their associated emissions factors. There might be other solid waste strategies that the team should consider, such as composting green waste and other organic matter on site or at another location. Each type of disposal for each type of material would have a different greenhouse gas emissions factor, which must be added to the transportation-related emissions.

That example illustrates four important points.

- When systems thinking is applied to sustainable design, it is often necessary to consider information beyond cost. A wide range of tools can help teams evaluate components of a system, including modeling, life cycle analysis, and life cycle cost analysis, as well as inventorying. These tools and technologies will be discussed in Section 4.

- Even if the system is evaluated using a complex computer model, the best solution may depend on the team's goals, metrics, and targets, as well as their resources. The alternatives must be analyzed and evaluated against the goals.

---

21   This is a hypothetical example and is not meant to imply that recycling costs more. In many cases, waste haulers with higher recycling rates charge lower fees because they have diversified their revenue streams.

- Although alternatives are often viewed as an either-or choice, there may be more than two options. In the waste hauler example, the question is about more than which hauler to select. When deciding between two alternatives, the project team must ask whether there is a third option (or a fourth or a fifth ...). The question can spark the creativity needed to find new solutions that lead to sustainability.

- Sometimes other variables, besides goals, targets, and costs, may make certain solutions inappropriate for the site. Sustainable design means finding not only the measures that perform best in a model but also the solutions that will perform best over the life of the project.

EVALUATING STRATEGIES

For existing building projects, the evaluation process should take the following steps:

- Set goals

- Benchmark performance

- Identify improvement opportunities

- Prioritize and align improvement opportunities with the project goals

- Implement the program

- Measure performance and undergo third-party verification

- Set revised or new goals

When a focus on performance requires the use of new technologies, sufficient time needs to be allotted for testing and inspections. The process of exploring and selecting technologies and strategies may be repeated as more information becomes available about the system. For example, in building energy analysis, modeling should be conducted very early in the project to inform initial decisions. As the project takes shape, the model is run again to evaluate general approaches to mechanical system design. The model might be refined when design development documents are 50% complete, and again at 75% and 90% of completion, to analyze the increasingly specific lighting systems, controls, and other components and strategies. In addition, modeling, design, and construction documents should be reviewed regularly by appropriate members of the project team, such as the commissioning agent. These commissioning reviews help ensure that the design meets the project goals defined at the beginning of the project.

As a project progresses, budget constraints often become apparent, and steps are needed to reduce costs. **Value engineering**, a formal review based on the project's intended function and conducted to identify alternatives that reduce costs and improve performance, is a critical part of the sustainable design process. Conceptually, this review fits in well with sustainable design, which is always focused on finding higher-performing, more efficient solutions. In practice, however, value engineering is often synonymous with cost cutting and is typically focused on first costs only; systems that have higher first costs but lower operating costs and higher efficiency may be abandoned. Any value engineering exercise must therefore keep the big picture in mind and include all stakeholders so that the decisions support the project goals.

IMPLEMENTATION

Once the planning and design phases are complete, it is time to think through each step of implementation and anticipate where problems might arise and compromise the project's commitment to sustainability. This up-front planning can help keep a project on schedule and on budget while protecting the project goals.

In both design-build and operations and maintenance projects, the first activities of the implementation phase focus on fine-tuning the decisions made during design and strategy selection, to make sure all selected strategies are practical given the constraints of construction.

FROM PLANNING TO PRACTICE

Management plans for design-build construction projects are critically important; they must be developed, implemented, and documented.

A **stormwater pollution prevention plan** addresses measures to prevent erosion, sedimentation, and discharges of potential pollutants to water bodies and wetlands.

An **indoor environmental quality management plan** spells out strategies to protect the quality of indoor air for workers and occupants; it includes isolating work areas to prevent contamination of occupied spaces, timing construction activities to minimize exposure to off-gassing, protecting the HVAC system from dust, selecting materials with minimal levels of toxicity, and thoroughly ventilating the building before occupancy.

A **waste management plan** addresses the sorting, collection, and disposal of waste generated during construction or renovation. It must address management of landfill waste as well as recyclable materials.

For operations and maintenance projects, the implementation phase may be less an event than an on-going process. Continual tweaks optimize operations, and major systems are overhauled for efficiency and ability to deliver energy and cost savings. Making sure everyone has the necessary training and information and clearly understands his or her role is the key to successful sustainable operations and maintenance programs.

With design-build projects, the construction process causes environmental damage, but the effects can be managed and reduced by using sound practices and alternative technologies.

The following strategies can help projects meet sustainability goals during construction:

- Reducing the amount of fossil fuels used in construction equipment by minimizing grading and earth moving, as well as using biodiesel or other alternative fuels.
- Preventing air and water pollution by addressing dust and implementing a **stormwater pollution prevention plan.**

- Ensuring indoor air quality by following an **indoor environmental quality management plan** for protecting ductwork and pervious materials, preventing dust, and protecting any occupied spaces from pollutants.
- Minimizing landfill waste by reducing construction debris and following a **waste management plan** that addresses waste separation and hauling, also saving costs.

As in all phases of a green building process, any changes made during implementation should be carefully documented. Although documentation may take time, it is necessary so that achievement of sustainability goals can be verified. Whether for compliance with regulatory requirements, LEED certification, or other third-party verification, clear and organized documentation throughout implementation will help ensure success. Documentation during the implementation phase might include change orders, chain-of-custody letters to verify that materials came from a sustainable source, waste hauling tickets, updated or revised construction management plans, commissioning or retrocommissioning reports, or other LEED documents. Careful recording and sharing of lessons learned can help improve future projects and advance the field of green building.

ON-GOING PERFORMANCE

The construction and operations of green building and neighborhood projects are never really complete. Daily life in any building or community requires on-going delivery or production of resources, as well as routine maintenance and upkeep. Even the most low-tech, passive systems need to be maintained to foster a healthful environment for people and prevent environmental harm to the planet. Heating, cooling, ventilation, and other systems must be properly cared for to ensure that they work effectively using minimum amounts of energy and water. Maintenance activities must be adapted throughout the life of the project so that the benefits are captured over time.

The key to understanding whether a project is performing sustainably is information—the right information at the right time. Data should document a project's on-going pursuit of sustainability goals. Project teams may be tempted to gather the data that are easy to collect and can be used as proof that the building is sustainable; the right data, in contrast, serve as honest, genuine feedback.

Orientation and training of the occupants and personnel must be repeated as new tenants move in, staff is hired, and lessons are learned. Education of building occupants encourages their full participation in sustainability opportunities. It helps stakeholders understand their role in optimizing performance and become vested in the green building goals. Education can take various forms, such as occupant luncheons, educational events, or interpretive signage. Tenant lease agreements, occupant handbooks, and staff training manuals will help newcomers benefit from a green project and contribute to its success. All members of the community should have easy access to information on how they can support sustainability and should be encouraged to participate and suggest improvements.

Just as with regular tune-ups and scheduled maintenance on an automobile, regular inspections and maintenance ensure that all building systems are performing well and continue to meet sustainability goals throughout the life of the project. Maintenance of mechanical, electrical, and plumbing systems is essential and needs to be included in regular operations budgets. Additional types of inspections to reveal problems or opportunities for improvement could include the following:

- Retrocommissioning
- Energy and water audits
- Solid waste audits
- Occupant surveys, including thermal comfort and transportation analysis
- Green purchasing and green housekeeping program assessments

These strategies will be discussed in Section 4.

On-going measurement and verification are essential to identifying opportunities for improvement. Sophisticated building automation systems are available to continuously collect and trend data; the process can also be conducted manually. The crucial next step is data analysis: a knowledgeable team member should regularly review the data, look for trends, spikes, or unusual values that may identify areas needing attention or repairs. Such observations can also reveal avenues to new energy and cost savings. Post-occupancy surveys complement performance-based data collection by indicating whether the project meets occupants' needs, is comfortable, and supports productivity.

The right information needs to flow to the right place. Whether that means measurement tools designed for daily use by maintenance staff, clear and accessible resource materials for occupants and residents, or collection and interpretation of building automation system outputs, the flow of information can be used as a feedback loop within the built environment to promote continuous improvements and support the commitment to sustainability.

Whether you are working on a small interior retrofit project or designing a whole new city, integrated sustainable design and operations processes support sustainability goals and innovation that lead to improvement.

> ### Success depends on these essentials:
> - Start early
> - Find the right team and process
> - Understand the systems across space and time
> - Develop clear and measurable goals
> - Follow an iterative process to ensure achievement of goals
> - Commit to continuous improvement

The next section will review specific concepts and strategies for different aspects of green design, planning, and operations. Each of these concepts and strategies should be viewed within the context of systems thinking, using integrated processes. This framework encourages green building practitioners to view projects as an interconnected system and thus find the best solutions for the built environment.

photo credit: Josh Partee 2009

ONE AND TWO POTOMAC YARD

The project at One and Two Potomac Yard certainly deserves recognition for its on-going performance efforts. From the beginning, this project established a commitment to long term sustainability as it pursued LEED for Existing Buildings certification, soon after achieving Gold Certification under the LEED for New Construction rating system. The facility management company, Jones Lang LaSalle and its Chief Building Engineer, Wayne DeGroat, rely on sophisticated feedback to ensure mechanical equipment is achieving maximum performance. As DeGroat points out, this ensures the project continues to meet performance targets. "We monitor everything through our EMS (environmental management system). It represents our whole mechanical plant...I can see water temperature, wet bulb readings, outside air humidity, and the outside air temperature." More information about the Potomac Yard projects is available at http://www.potomacyard.net/.

SECTION 4
GREEN BUILDING CORE CONCEPTS AND APPLICATION STRATEGIES

The first sections of this guide set a foundation for green building practice by addressing integrated processes and the reasons for departure from conventional practice. This section builds on that groundwork, presenting fundamental concepts alongside strategies for putting green building into action. It is critical to keep a systems approach, as discussed in section 2, in mind; however, categorization of concepts and strategies provides a framework for application.

Although there are many ways to organize green building projects, this section follows the general categories associated with the LEED rating systems (whose specific categories and titles vary):

- **Sustainable Sites and Location**
- **Water Efficiency**
- **Energy and Atmosphere**
- **Materials and Resources**
- **Indoor Environmental Quality**
- **Innovation**

Despite this organizational framework, many synergistic opportunities can be found both within and between categories. For example, the location of a project can have a significant effect on occupants' transportation choices, the project's energy needs, and potential opportunities for using renewable energy.

SUSTAINABLE SITES

The location of a building is as important as how it is built. Its connection and linkage to the local bioregion, watershed, and community will help determine how a project can contribute to a sustainable environment. A sustainable project serves more than the immediate function of the building. It must also meet the needs of the local community, support active street life, promote healthy lifestyles, provide ecosystem services, and create a sense of place. Site selection and design play important roles in both reducing greenhouse gas emissions and helping projects adapt to the effects of climate change. If people can use public transportation, ride bicycles, or walk to the building, the project helps reduce the carbon emissions associated with commuting. A project that is connected to the community by pedestrian paths and bicycle lanes encourages people to walk or bike instead of drive, not only helping to reduce air pollution, but also promoting physical activity.

In ideal cases, sustainable design projects start in one of two ways—either the team starts with a site and considers the best functions and uses for that particular location, or the team starts with a function and determines the best location for that land use. In either case, by understanding the needs and goals of the project as well as the opportunities and constraints of a particular location, the team will be able to arrive at an optimal set of solutions.

When selecting a site, the team must consider many attributes of the overall system:

- What is the local climate of the project?
- Has the site been previously developed?
- Is it connected to local infrastructure and public transportation?
- What species in the area might use the site as habitat and be affected?
- What is the nature of the street life in the area, and how can the project contribute to community?
- Where do people in the area live and work, and how do they get back and forth?

LEED rating systems address project location and site design and maintenance through the following topics:

- Location and linkage
- Neighborhood pattern and design
- Transportation
- Stormwater management
- Heat island effect

LOCATION AND LINKAGE

A good project site channels development into places where it will improve, rather than degrade, the triple bottom line. The best locations are those that promote **smart growth**, an approach that protects open space and farmland by emphasizing development with housing and transportation choices near jobs, shops, and schools. **Infill development** uses sites that have been previously developed or are gaps between existing structures. This practice helps limit the amount of land covered by buildings, pavement, or infrastructure while also better using the space within existing communities.

Brownfield sites, in particular, can actually improve environmental performance. The U.S. Environmental Protection Agency (EPA) defines **brownfields** as land where development may be complicated by the presence or potential presence of hazardous substances, pollutants, or contaminants.[22] Development or redevelopment of brownfields may require the cleanup of contaminated soil or groundwater. Brownfields provide real opportunities for green building projects to go beyond just reducing their effects on the environment and enhance the community.

---

22 U.S. Environmental Protection Agency, *Brownfields Definition* (2009), http://epa.gov/brownfields/overview/glossary.htm.

Other sites are less appropriate for development. For example, development of sites that have been in agricultural use, called **greenfields**, and sites that are far from existing development and infrastructure will increase the total development footprint, reduce the amount of land available for open space or agriculture, and fragment wildlife habitat; it may encourage development to continue outside urban boundaries. Development is also discouraged in wetland areas, floodplains, steep slopes, and endangered species habitat.

Strategies to address location and linkages:

- **Choose redevelopment and infill development.** Build on previously developed land and brownfield sites.
- **Locate near existing infrastructure.** Avoid triggering suburban sprawl and unnecessary materials use by consolidating development along existing roads, power lines, and water supplies.
- **Protect habitat.** Give preference to locations that do not include sensitive site elements and land types.
- **Increase density.** Create a smaller footprint and maximize the floor-area ratio or square footage per acre.
- **Increase diversity.** Provide the services that are most needed within communities and support a balance of jobs and housing.
- **Encourage multiple modes of transportation.** Enable occupants to walk, bicycle, and use public transit.

NEIGHBORHOOD PATTERN AND DESIGN

Community layout and planning influence occupants' and residents' behavior while setting a standard for what these locations value. For example, where culs-de-sac connect to increasingly wide connector roads, services are clustered into strip malls, and jobs are centered in office parks, the emphasis is on the private realm and the automobile. On the other hand, in communities with connected street grids, diverse land uses, and buildings facing wide sidewalks, the emphasis is on pedestrians and the public realm. Neighborhood pattern and design strategies are those that help make a project easy to navigate, accessible, and appealing to pedestrians. The focus is on the diversity of land uses, the design of streets, and the functions of the community. Residents meet their needs within their neighborhoods, including going to work or school, finding places to meet or play, and getting healthful food.

Strategies for sustainable neighborhood pattern and design:

- **Design walkable streets.** Focus on building frontage, ground-level façade, building height-to-street-width ratio, and sidewalks. Limit street speeds. Include street trees, shade, benches, and other amenities for pedestrians.
- **Use compact development strategies.** Consolidate development by increasing the number of units of residential space and square feet of commercial space per acre.

- **Promote connectivity.** Limit culs-de-sac, prohibit gated communities, and use a street grid pattern.

- **Provide diverse land uses.** Include a wide mix of services, such as shops, restaurants, schools, religious centers, grocery stores, parks, civic buildings, and recreational facilities.

- **Create a diverse community.** Provide housing types for a wide range of incomes and abilities. Incorporate, rather than segregate, affordable and senior housing.

- **Promote alternative transportation.** Limit parking, connect the buildings to public transit and bicycle paths, and provide transit centers.

- **Support access to sustainable food.** Include community gardens, farmers markets, urban farms, and community-supported agriculture programs. Ensure that all residents have easy access to grocery stores and other food choices beyond fast food.

TRANSPORTATION

According to the U.S. Energy Information Administration, transportation accounted for 33% of total U.S. greenhouse gas emissions in 2008.[23] Globally, transportation is responsible for 13.5% of total carbon dioxide emissions.[24] Generally, this is a result of three fundamental factors: land use, vehicle technology, and transportation fuels.

Attention to each of those factors can reduce the consequences of transportation. Land-use decisions can help reduce the length and frequency of vehicle trips and encourage shifts to more sustainable modes of transportation. Vehicle technology determines the quantity and types of energy and support systems needed to convey people and goods to and from the site. Fuel determines the environmental effect of vehicle operation. Current efforts to improve vehicle fuel efficiency and reduce the carbon intensity of motor fuels may be insufficient to meet greenhouse gas reduction goals unless accompanied by significant changes in land use and human behavior. Regardless of substantial investments in technology and alternative energy, poor planning can still cause a net increase in greenhouse gas emissions as commuters weigh options for how they travel to and from work, school, home, and errands.

23 U.S. Energy Information Administration, *Emissions of Greenhouse Gases Report* (December 8, 2009), http://www.eia.doe.gov/oiaf/1605/ggrpt/.

24 K. Baumert, T. Herzog, and J. Pershing, *Navigating the Numbers: Greenhouse Gas Data and International Climate Policy* (Washington, D.C.: World Resources Institute, 2005).

Promoting alternative transportation as a convenient and viable option through site selection, design, and incentives benefits both the building occupants and the developer. The LEED rating systems give project teams flexibility when considering site-specific needs and opportunities for alternative transportation. Project teams can reduce transportation effects by ensuring access to alternative modes of transportation, encouraging walking and bicycling, and providing fueling facilities for alternative-fuel vehicles. Project teams are also rewarded for reducing the number and length of automobile trips by locating in high-density areas or infill sites already served by mass transit, where occupants and visitors can more easily use existing transportation networks. Sites without access to public transportation start at a disadvantage and may require additional attention to transportation, particularly local land-use design and alternative fuels. It is still possible for such a project to substantially reduce its transportation effects if the team focuses on local connectivity and the energy efficiency of the vehicles used to serve its needs. For example, an office complex without transit access might provide incentives for carpooling, incorporate diverse land uses that allow workers to walk to basic services, or facilitate the use of alternative-fuel vehicles like plug-in hybrids.

Strategies to address transportation in design and planning:

- **Locate near public transit.** Select a project site in an area served by an existing transportation network.
- **Limit parking.** The lack of parking spaces on the project site will spark interest in alternative transportation options.
- **Encourage bicycling.** Install secure bike racks and showers for commuters.

Strategies to address transportation in operations and maintenance:

- **Encourage carpooling.** Designate preferred spaces for carpool vehicles in the parking area.
- **Promote alternative-fuel vehicles.** Provide a convenient refueling station on the site.
- **Offer incentives.** Develop an alternative commuting incentive program for building occupants.
- **Support alternative transportation.** Promote alternatives to single-occupancy car commuting at the building and/or city level.

LEED IN PRACTICE
Smart locations and LEED for Neighborhood Development

LEED for Neighborhood Development encourages development within and near existing communities or public transportation infrastructure. The goal is to reduce vehicle trips and miles traveled and support walking as a transportation choice. This promotes public health and a vibrant community life.

One measure of "smart location" is access to transit service. LEED recognizes projects that locate dwelling unit entrances within a certain walking distance of bus and streetcar stops, rapid transit, light or heavy passenger rail stations, ferry terminals, or tram terminals.

Figure 4.1. Illustration of the evaluation of a ½ mi walk distance to public transportation – one measure of a "Smart Location". (Source: LEED Reference Guide for Green for Neighborhood Development, Washington, DC, 2009)

SITE DESIGN AND MANAGEMENT

Projects may set broad goals for sustainable design and management of a site, such as reducing the environmental impacts of landscaping, minimizing maintenance costs, and contributing to the restoration and regeneration of an area. Achieving these goals requires careful plant selection, integration of innovative irrigation systems, and a new approach to outdoor lighting design.

Strategies for designing and maintaining a sustainable site can include selecting native and adapted species that thrive without irrigation, pesticides, or fertilizers. Certain plants can enhance soil nutrients, supporting regenerative project goals; others naturally deter pests. Plants can also be selected to minimize **evapotranspiration**, the return of water to the atmosphere through evaporation from plants' leaves; this characteristic is important in arid climates. Strategic selection of plants creates wildlife habitat and support **integrated pest management (IPM)**, a sustainable approach that combines knowledge about pests, the environment, and pest prevention and control methods to minimize pest infestation and damage in an economical way while minimizing hazards to people, property, and the environment.

Strategically locating functional and decorative hardscape on a project site may reduce the amount of **impervious area**, surfaces that have been compacted or covered by materials that do not allow water to infiltrate. Impervious areas found in the built environment include concrete, brick, stone, asphalt, and sealed surfaces. Strategies for reducing hardscape include using pervious paving systems for parking lots, walkways, and decorative areas, such as patios. Pervious paving areas allow stormwater infiltration and also reduce heat island effects.

The benefits of sustainable site design and management reach far beyond a project's boundary. Site lighting can provide adequate nighttime illumination while preserving the integrity of the night sky. By reducing glare and contrast between light and dark areas, which can diminish night vision, smart lighting design can actually improve site safety while maintaining views of the stars and decreasing stress to nocturnal animals. To achieve such goals, teams avoid up-lighting and over-lighting, direct full cutoff fixtures downward to illuminate paths and exits, and shield fixtures to prevent **light trespass**, the spilling of light beyond the project boundary. In the evening, reflective paving materials help distribute light across the site, reducing the number of fixtures needed to safely illuminate the area while also saving energy. Where higher light levels are needed, timers shut off lighting late at night.

Strategies for developing a sustainable site design:

- **Minimize hardscape.** Design driveways and paths intelligently. Substitute pervious surfaces for traditional paving.

- **Use native landscaping.** Select plants that are native to the area both to reduce water use and to provide habitat for local birds and other species. Incorporate mulch into the landscape to build the soil and naturally suppress weeds.

- **Prevent light pollution.** Avoid up-lighting, glare, and trespass by using shielded fixtures and smart lighting design.

- **Preserve open space and sensitive areas.** Consolidate the development footprint and protect and restore natural vegetation, wetland areas, and bodies of water.

- **Protect and restore habitat.** Designate areas as protected habitat and open space for the life of the project. Develop a conservation management program to make sure that the natural environment is protected. Consider putting protected areas into a land trust.

Strategies for sustainable site operations and maintenance:

- **Develop a sustainable site management plan.** The plan should address the application of chemicals and the cleaning of hardscape and building exterior, and it should include an integrated pest management program.

- **Implement conservation programs.** Work with ecologists and nonprofit organizations to implement conservation programs that protect species and habitat.

- **Maintain site lighting to prevent light pollution.** Ensure that fixtures are replaced according to the original design. If higher light levels are needed, include timers that shut them off automatically after hours.

STORMWATER MANAGEMENT

The stormwater systems of most American urban areas treat precipitation as a problem to be removed from the area as quickly as possible to prevent flooding. The result, combined with the ever-expanding boundary of the urban edge and the increase in paved roads and hardscape, is damaging to the watershed function. The alternative, applying systems-based, integrated processes to stormwater management, encourages teams to mirror natural systems by slowing the flow of water and emphasizing on-site water retention. They can increase infiltration of rainfall into the ground, capture and reuse it, and use natural processes to treat the remaining water that runs off the property.

Impervious surfaces, such as asphalt and concrete, prevent percolation and infiltration and encourage water runoff, causing soil erosion and in some places sedimentation of local waterways. This runoff can also carry harmful chemicals into the water system, degrading surface water quality and harming aquatic life and recreation opportunities in receiving waters. This **nonpoint source pollution**, from diffuse land uses rather than a single facility, is one of the biggest threats to surface water quality and aquatic ecosystems.

LEED recognizes and encourages planning, design, and operational practices that control stormwater and protect the quality of surface and ground water. Many of these solutions fall within the scope of **low-impact development (LID)**, an approach to land management that mimics natural systems and manages stormwater as close to the source as possible.[25] It includes minimizing impervious surfaces, protecting soils, and enhancing native vegetation. The Department of Environmental Resources in Prince George's County, Maryland, for example, uses LID control measures that integrate five components: site planning, hydrologic analysis, integrated management practices, erosion and sediment control, and public outreach. This approach protects surface water by managing stormwater on site and creating buffers between development and water resources.

Stormwater management can also include the collection and reuse of water for nonpotable purposes, such as landscape irrigation, toilet and urinal flushing, and custodial uses. This helps reduce stormwater runoff while avoiding the unnecessary consumption of expensive and energy-intensive potable water. The strategy illustrates the importance of understanding a region's environmental conditions. For example, in the eastern United States, on-site water collection is often encouraged as part of efforts to slow stormwater runoff and reduce nonpoint source pollution. Conversely, in some western states, long-standing water laws prohibit on-site water collection because the water is obligated to downstream users.

25 U.S. Environmental Protection Agency, *Low Impact Development* (2011), http://www.epa.gov/owow/NPS/lid/.

HEAT ISLAND EFFECT

Cities are typically warmer than nearby rural areas. The flat, dark surfaces of roadways, parking lots, and tarred rooftops absorb and retain the sun's heat during the day and are slow to radiate it at night. The result, known as the **heat island effect**, is an increase in air temperature in a developed area compared with an undeveloped area. The increased heat absorption in urban areas has several consequences:

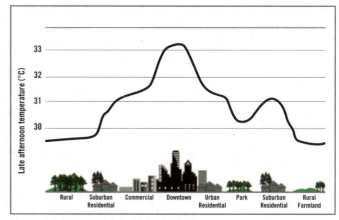

Figure 4.2. Diagram of Heat Island Effect

- The additional use of air-conditioning increases energy demand and costs. The rise in energy costs is dramatic because the highest demand for air-conditioning occurs during peak hours for energy consumption.
- Populations of wildlife species not adapted to the higher temperature (and its effects on the environment, including changes in resource availability) decline.
- Wildlife species not adapted to the higher temperature (and its effects on the environment, including changes in resource availability) decline.

To mitigate those harmful effects, project teams can install surfaces that have high **albedo** or a high **solar reflectivity index (SRI)**. Albedo is a reflectivity measurement. SRI combines reflectivity with **emissivity**, or the ability of a material to emit energy through radiation. The use of reflective materials and those with high SRI values reduces heat gain, thus increasing comfort and reducing demand for air-conditioning. Materials that help reduce the heat island effect include concrete paving (instead of asphalt), white roofs, and vegetated "green" roofs.

Strategies for reducing the heat island effect:

- **Install reflective roof surfaces.** Light-colored or white roofs absorb less heat.

- **Reduce the area of paved surfaces exposed to sunlight.** Limit the amount of hardscape, design narrow roads, use light-colored paving, shade hardscape with greenery, locate parking underground.

- **Plant an urban forest or a green roof.** Use street trees, shrubs, and landscaping to reduce heat island effects through evapotranspiration as well as shade.

WATER EFFICIENCY

The U.S. Geological Survey estimates that the United States uses more than 400 billion gallons of water per day. The operation of buildings, including landscaping, accounts for approximately 47 billion gallons per day—12% of total water use.[26] As residential, commercial, industrial, and other development expands, so does the use of the limited **potable water** supply, water that is suitable for drinking. Most buildings rely on municipal sources of potable water to meet their needs, from flushing toilets to washing dishes and landscape irrigation. High demand strains supplies and under extreme conditions necessitates water rationing. Furthermore, large amounts of wastewater can overwhelm treatment facilities, and the untreated overflow can contaminate rivers, lakes, and the water table with bacteria, nitrogen, toxic metals, and other pollutants. To avoid this damage to the ecosystem, additional municipal supply and treatment facilities must be built, at public cost. Water pumping and treatment, both to and away from the project, also require energy, whose production generates additional greenhouse gas emissions.

26 S.S. Hutson, N.L. Barber, J.F. Kenny, K.S. Linsey, D.S. Lumia, and M.A. Maupin, *Estimated Use of Water in the United States in 2000* (2004), http://pubs.usgs.gov/circ/2004/circ1268/pdf/circular1268.pdf.

Green building encourages innovative water-saving strategies that help projects use water wisely. Project teams can follow an integrated process to begin assessing existing water resources, opportunities for reducing water demand, and alternative water supplies. For example, much of the water that leaves the site as waste water or stormwater runoff can actually be used for nonpotable functions.

Guiding questions for a team to consider during this process may include the following:

- How much rain falls on the site per year?
- How will water be used on site, and how can the amount be reduced?
- What are the sources of **greywater**, such as from sinks and showers, that can easily be collected and reused for nonpotable uses, such as irrigation?

Some project teams use their sites' annual precipitation to determine how much water they should use. Clearly, the **water balance** approach is more achievable for projects that receive more rain and require less irrigation. However, projects around the country are experimenting with this goal. It requires reducing demand by designing sites to minimize or eliminate the need for irrigation and installing plumbing fixtures that either conserve water (such as low-flow lavatories and dual-flush toilets) or eliminate demand entirely (such as waterless urinals and composting toilets). Additionally, captured stormwater and treated greywater can be used instead of potable water for toilet flushing, irrigation, and cooling towers.

The value of any particular measure for overall water conservation efforts depends on the end uses in the project. For example, office buildings typically lack extensive laundry and kitchen facilities; water is used for HVAC systems, restrooms, and landscaping. In contrast, kitchen sinks and dishwashers dominate the end use for restaurants. A water end-use profile can help project teams identify the largest users of water and evaluate the cost-effectiveness of specific conservation strategies, whether low-flow fixtures, irrigation technology, or efficient cooling tower systems.

Efficiency strategies, combined with monitoring systems that track water consumption and identify problems as they arise, can dramatically improve water conservation compared to conventional building water use. LEED rewards projects that both reduce demand and reuse water for indoor and outdoor water uses.

INDOOR WATER USE

Indoor use encompasses water for urinals, toilets, showers, kitchen or break room sinks, and other applications typical of occupied buildings. Indoor water use can be reduced by installing water-efficient fittings and fixtures, using nonpotable water for flush functions, and installing submeters to track and log water use trends, check fixture performance, and identify problems. Buildings also use significant amounts of water to support industrial processes and systems, such as cooling towers, boilers, and chillers. These systems provide

both heat and cool air and water for building operations. Process water also includes the water used for certain business operations (e.g., washing machines, dishwashers). Commercial building projects can reduce water use by selecting efficient cooling towers, chillers, boilers, and other equipment, and by substituting harvested rainwater and nonpotable water for certain applications.

Understanding how water is being used allows teams to identify where they should focus conservation efforts. Submeters report how much water is being used by systems and fixtures and alerts managers to leaks or other inefficiencies. Metering the water lost to evaporation during cooling tower operation can provide particularly important information. Facilities may be able to receive credit from the utility company for sewer charges if they reduce the amount of water entering the sewer system.

Strategies for reducing indoor water use:

- **Install efficient plumbing fixtures.** Install new low-flow fixtures, including low-flow lavatories, kitchen sinks and showers, dual-flush toilets, waterless urinals, and composting toilets. Low-flow fixtures use less water than specified by the Energy Policy Act (EPAct) of 1992. Select EPA WaterSense and EnergyStar products. In existing buildings, if porcelain replacement proves cost-prohibitive, install new flush valves or flow restrictors (e.g., aerators) to achieve water savings.

- **Use nonpotable water.** If permitted by the jurisdiction, use captured rainwater, greywater, or municipally provided reclaimed water for flush fixtures. Design and install plumbing systems that can use captured rainwater or greywater in flush fixtures. Greywater use is not an option in all municipalities, so it is important check regulations before planning to utilize this strategy.

- **Install submeters.** Meter indoor water systems and monitor the data to track consumption trends, determine fixture performance, and pinpoint leaks.

OUTDOOR WATER USE

Landscape irrigation, a significant component of many commercial buildings' water use, presents an important opportunity to conserve water. Reductions in irrigation can be achieved by specifying water-wise landscaping and water-efficient irrigation technology, using nonpotable water, and installing submeters to track and log irrigation trends.

Native and adapted species support water efficiency goals because these plants typically don't need to be irrigated. **Xeriscaping** is the use of drought-tolerant native or adapted plants along with rocks, bark mulch, and other landscape elements. High-performance irrigation systems, such as drip systems and bubbler distribution systems, channel water directly to root systems; weather-based irrigation controllers respond to weather conditions. Potable water use for irrigation can be further reduced by using nonpotable water for outdoor applications.

Finally, as with indoor water use, submetering helps teams understand how much water is being used for irrigation.

Strategies for reducing outdoor water use:

- **Choose locally adapted plants.** Landscape with native and adapted plants that require less water. These plantings have the added benefit of providing habitat for native wildlife.

- **Use xeriscaping.** These drought-tolerant plantings have extremely low water needs. Especially in arid regions, employ xeriscape principles when designing the site landscape.

- **Select efficient irrigation technologies.** Drip and bubbler systems and weather-based controllers can save water.

- **Use nonpotable water.** Captured rainwater, greywater, or municipal reclaimed water is suitable for irrigation.

- **Install submeters.** Meter the irrigation system to track water consumption and identify leaks.

ENERGY AND ATMOSPHERE

Energy has emerged as a critical economic issue and top priority for policymakers. Unsustainable energy supply and demand have serious implications for everything from household budgets to international relations. Buildings are on the front line of this issue because of their high consumption of energy. Studies have repeatedly shown that efficient buildings and appropriate land use offer opportunities to save money while reducing greenhouse gas emissions. One such study, conducted by the New Buildings Institute, investigated 121 LEED-certified commercial office buildings in the United States and found that they used 24% less energy than the national average. Almost half of the buildings in the study achieved an **ENERGY STAR Portfolio Manager** score of 75 or above, with an overall average score of 68.[27]

27 C. Turner and M. Frankel, *Energy Performance of LEED for New Construction Buildings* (March 4, 2008), http://www.newbuildings. org/sites/default/files/Energy_Performance_of_LEED-NC_Buildings-Final_3-4-08b.pdf.

Set up by EPA as a part of the ENERGY STAR program, **ENERGY STAR Portfolio Manager** is an interactive, online management tool that supports tracking and assessment of energy and water consumption. In Portfolio Manager, a score of 50 represents average building performance. The New Buildings Institute study also collected data suggesting that a significant percentage of buildings underperformed their benchmarks. This finding reinforces the importance of commissioning systems and monitoring performance so that green buildings can maintain their efficiencies and achieve their full potential over time.

The design and operations of buildings, neighborhoods, and communities can dramatically boost energy efficiency and the benefits from cleaner, renewable energy supplies.

> **Following an integrated process helps identify synergistic strategies for the following areas:**
>
> - Energy demand
> - Energy efficiency
> - Renewable energy
> - Ongoing performance

ENERGY DEMAND

Saving energy begins with conservation—reducing energy demand. Green buildings and neighborhoods can reduce demand for energy by capturing natural, incident energy, such as sunlight, wind, and geothermal potential, to reduce loads. For example:

- Community planning can support building configurations that minimize solar gain in summer and maximize it in winter
- Adjacent buildings can be designed to shade and insulate each other
- Building designs that incorporate passive strategies, like daylight, thermal mass, and natural ventilation, reduce the demand for artificial lighting, heating, and cooling
- Technologies and processes can be used to help occupants understand their patterns of energy consumption and reduce both individual and aggregate energy demand

In addition to reducing demand, green building encourages sustainable methods for meeting energy needs. This may be most applicable when addressing a project's use of **refrigerants**, substances used in cooling of systems. Refrigerants were widely employed throughout the 20th century for transferring thermal energy in air-conditioning and refrigeration systems. Although these substances have remarkable functional properties, they also have damaging side effects on the environment. In the 1980s, research emerged demonstrating that certain refrigerants for building systems were depleting stratospheric ozone, a gas that protects human health and the environment by absorbing harmful UV radiation, and contributing to climate change. The **Montreal Protocol** subsequently banned the production of **chlorofluorocarbon (CFC)** refrigerants and is phasing out **hydrochlorofluorocarbon (HCFC)** refrigerants. CFCs and HCFCs are organic chemical compounds known to have ozone-depleting potential.

To achieve LEED certification, new buildings may not use CFC-based refrigerants, and existing buildings must complete a total CFC phase-out prior to project completion. LEED awards points for projects that entirely avoid the use of refrigerants or select refrigerants that balance concerns about ozone depletion and climate change. LEED recognizes that although there are no perfect refrigerants, it is possible to carefully consider performance characteristics and environmental effects and select a refrigerant with an acceptable trade-off.

Taken together, demand reduction strategies provide the foundation for further energy efficiency efforts and the effective use of renewable energy.

Strategies for reducing energy demand in design and planning:

- **Establish design and energy goals.** Set targets and establish performance indicators at the outset of a project and periodically verify their achievement.
- **Size the building appropriately.** A facility that is larger than necessary to serve its function creates costly and wasteful energy demand.
- **Use free energy.** Orient the facility to benefit from natural ventilation, solar energy, and daylight.
- **Insulate.** Design the building envelope to insulate efficiently against heating and cooling losses.

Strategies for reducing energy demand in operations and maintenance:

- **Use free energy.** Use the facility's orientation and appropriate shades, windows, and vents to take advantage of natural ventilation, solar energy, and daylight.
- **Monitor consumption.** Use energy monitoring and feedback systems to encourage occupants to reduce energy demand.

LEED IN PRACTICE
Reduce demand by reducing building size.

Energy demand typically increases in direct relation to building size: the more square feet in a building, the more energy it consumes. Although there are exceptions, the relationship between square footage and consumption is very strong.

The LEED for Homes rating system includes an adjustment to compensate for

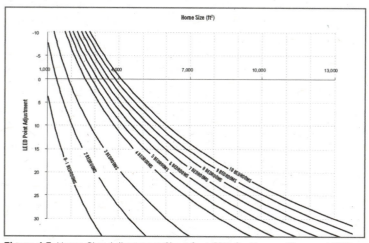

Figure 4.3. Home Size Adjustment Chart for LEED for Homes (Source: LEED for Homes Reference Guide, Second Edition 2009, Washington, DC, 2009.)

the effect of square footage on resource consumption by adjusting the point thresholds for Certified, Silver, Gold, and Platinum ratings based on home size (Figure 4.3). The adjustment applies to all LEED for Homes credits, not just to strategies related to Energy and Atmosphere.

The adjustment explicitly accounts for the material and energy impacts of home construction and operation. Depending on design, location, and occupants' behavior, a 100% increase in home size yields an increase in annual energy use of 15% to 50% and an increase in materials usage of 40% to 90%.

LEED for Homes is currently the only LEED rating system with this type of adjustment.

ENERGY EFFICIENCY

Once demand reduction strategies have been addressed and incorporated, the project team can begin to employ strategies to promote **energy efficiency**—using less energy to accomplish the same amount of work. Getting the most work per unit of energy is often described as a measure of energy intensity. Common metrics for buildings and neighborhoods include energy use per square foot and use per capita. Figure 4.2 outlines a typical office building's energy use. Each category provides an opportunity for increasing efficiency and savings.

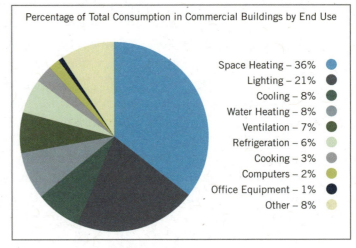

Percentage of Total Consumption in Commercial Buildings by End Use

- Space Heating – 36%
- Lighting – 21%
- Cooling – 8%
- Water Heating – 8%
- Ventilation – 7%
- Refrigeration – 6%
- Cooking – 3%
- Computers – 2%
- Office Equipment – 1%
- Other – 8%

Figure 4.4. Distribution of Building Energy Use

Through the integrated process, green building project teams can identify opportunities for employing synergistic strategies. For example, by improving the **building envelope**, the space between exterior and interior environments of a building which typically includes windows, walls, and roof, teams may be able to reduce the size of HVAC systems or even eliminate them altogether. This kind of integrated design can reduce both initial capital costs and long-term operating costs.

Strategies for achieving energy efficiency:

- **Address the envelope.** Use the regionally appropriate amount of insulation in the walls and roof and install high-performance glazing to minimize unwanted heat gain or loss. Make sure that the building is properly weatherized.

- **Install high-performance mechanical systems and appliances.** Apply life cycle assessment to the trade-offs between capital and operating costs, and evaluate investments in energy efficiency technologies. Appliances that meet or exceed ENERGY STAR requirements will reduce plug load demands.

- **Use high-efficiency infrastructure.** Efficient street lighting and LED traffic signals will reduce energy demands from neighborhood infrastructure.
- **Capture efficiencies of scale.** Design district heating and cooling systems, in which multiple buildings are part of a single loop.
- **Use energy simulation.** Computer modeling can identify and prioritize energy efficiency opportunities.
- **Monitor and verify performance.** Ensure that the building systems are functioning as designed and support the owner's project requirements through control systems, a building automation system, and commissioning and retrocommissioning.

RENEWABLE ENERGY

Reduced demand and increased efficiency often make it cost-effective to meet most or all of a building's energy needs from renewable sources. So-called **green power** is typically understood to include solar, wind, wave, biomass, and geothermal power, plus certain forms of hydropower. Use of these energy sources avoids the myriad of environmental impacts associated with the production and consumption of non renewable fuels, such as coal, nuclear power, oil, and natural gas.

LEED distinguishes between on-site renewable energy production and purchase of off-site green power. On-site energy production typically involves a system that generates clean electricity, such as solar photovoltaic panels that convert the sun's energy into electricity. Off-site renewable energy is typically purchased at a premium price per kilowatt-hour from a utility or a provider of **renewable energy certificates (RECs)**. RECs represent a tradable, nontangible commodity associated with the qualities of renewable electricity generation. RECs, and their associated attributes and benefits, can be sold separate from the underlying physical electricity associated with a renewable-based generation source. A project team that cannot purchase green power through the local utility can offset the building's energy use by purchasing green power from renewable energy projects around the country. Sometimes project teams can enter into REC agreements that provide for specific energy sources, such as wind or biomass, from a particular generation facility.

Strategies for meeting energy demand with renewable energy:

- **Generate on-site renewable energy.** Install photovoltaic cells, solar hot water heaters, or building-mounted wind turbines.
- **Purchase off-site renewable energy.** Buy green power or renewable energy certificates to reduce the environmental impact of purchased electricity and promote renewable energy generation.

ONGOING ENERGY PERFORMANCE

Attention to energy use does not end with the design and construction of an energy-efficient building. It is critical to ensure that a project functions as designed and that it sustains and improves this performance over time. Performance goals set during planning and design can be undermined by design flaws, construction defects, equipment malfunctions, and deferred maintenance. Monitoring and verification provide the basis for tracking energy performance, with the goal of identifying and resolving any problems that may arise. Monitoring often involves comparing building performance measurements with predictions from a calibrated energy simulation or industry benchmarking tool. EPA's ENERGY STAR Portfolio Manager is one of the most widely used benchmarking systems. Users enter data on electricity and natural gas consumption, along with other supporting information, into a Web-based tool. The system then evaluates the performance of the building against that of others with similar characteristics. This is an exceptionally useful, free tool for gauging the relative performance of buildings.

Commissioning is a systematic investigation by skilled professionals who compare building performance with performance goals, design specifications, and most importantly, the owner's requirements. This process begins early in design, with the specification of requirements. The requirements are considered throughout the building design and construction process and become the baseline for evaluation. Ongoing commissioning for building operations ensures that a building continues to meet its fundamental operational requirements. **Retrocommissioning** is the same process applied to existing buildings; it is intended to keep a building on track for meeting or exceeding the original operational goals.

The cost of commissioning is often repaid with recovered energy performance. A Lawrence Berkeley National Laboratory study found that commissioning for existing buildings had a median cost of $0.27 per square foot and yielded whole-building energy savings of 15%, with an average simple payback period of 0.7 years. For new construction, median cost was determined to be $1 per square foot with a median payback time of 4.8 years based on energy savings alone.[28] Overall, this study concluded, commissioning is one of the most cost-effective means of improving energy efficiency in commercial buildings.

LEED recognizes and encourages operational energy performance through its requirements for building commissioning and credits for monitoring and verification.

> **Strategies for incorporating ongoing performance measurement into a project:**
>
> ● **Adhere to the owner's project requirements.** Prepare detailed owner's project requirements at the beginning of the design process and conduct commissioning throughout the life cycle of the project to ensure that the building functions as designed.

28 E. Mills et al., *The Cost Effectiveness of Commercial Buildings Commissioning: A Meta-Analysis of Existing Buildings and New Construction in the United States* (November 23, 2004), http://www.dot.ca.gov/hq/energy/Cx-Costs-Benefits.pdf.

- **Provide staff training.** Knowledge and training empower facilities managers to maintain and improve the performance of buildings.

- **Conduct preventive maintenance.** Develop a robust preventive maintenance program to keep the building in optimal condition.

- **Create incentives for occupants and tenants.** Involve building occupants in energy efficiency strategies. Promote the use of energy-efficient computers and equipment, bill tenants from submeter readings to encourage energy conservation, educate occupants about shutting down computers and turning out lights before they leave, and give them regular feedback on energy performance.

MATERIALS AND RESOURCES

Materials and resources are the foundation of the buildings in which we live and work, as well as that with which we fill them, the infrastructure that carries people to and from these buildings, and the activities that take place within them. The ubiquitous nature of materials and resources makes it easy to overlook the history and costs associated with production, transportation, consumption, and disposal. The "Story of Stuff," as this process has become known from the popular YouTube video and subsequent book by the same name, often begins as raw materials from around the world. They are transported, refined, manufactured, and packaged for sale. In a conventional system, stuff is purchased, consumed, and discarded, often in a landfill. But in reality, there is no "away" and each step in this process of production, consumption, and disposal has significant environmental, social, and economic consequences.

Setting goals for using sustainable materials and resources is an important step of the green building process. "Reduce, reuse, recycle" may seem like a critical component of this work: clearly, reducing consumption is critical, and reusing and recycling waste are important strategies. But green building requires rethinking the selection of materials as well. Ideally, the materials and resources used for buildings not only do less harm but go further and regenerate the natural and social environments from which they originate. To evaluate the best options and weigh the trade-offs associated with a selection, teams must think beyond a project's physical and temporal boundaries. Life cycle assessment can help a team make informed, defensible decisions.

Plentiful opportunities exist to reduce the harms associated with materials. Using less, finding materials with environmentally preferable attributes, using locally harvested materials, and eliminating waste provide a great starting place. A systems-based, life cycle perspective and an integrative process will help projects achieve their goals addressing materials and resource use. LEED addresses the following issues related to materials and resources:

- Conservation of materials
- Environmentally preferable materials
- Waste management and reduction

CONSERVATION OF MATERIALS

A building generates a large amount of waste throughout its life cycle. Meaningful waste reduction begins with eliminating the need for materials during the planning and design phases. For example, compared with sprawling communities, denser, more compact mixed-use urban neighborhoods require fewer miles of road and less physical infrastructure to support the same number of people. Similarly, smaller, more efficiently built buildings and homes require fewer board-feet of lumber or linear feet of pipe, as well as fewer resources to maintain. Experienced contractors often have great ideas for implementing such material-saving strategies. Bringing them in at the early phases of an integrated process, instead of waiting until the design is complete, can add real value to the design team and the project as a whole.

Materials procurement doesn't end at the end of construction. For example, companies' on-going procurement strategies can provide real opportunities to reduce material usage.

Strategies for conserving materials throughout a project's life cycle:

- **Reuse existing buildings and salvaged materials.** Selecting resources that have already been harvested and manufactured results in tremendous materials savings.
- **Plan for smaller, more compact communities.** Reduce the need for new roads and other infrastructure by preventing sprawling land-use patterns.

ENVIRONMENTALLY PREFERABLE MATERIALS

Many attributes can be the basis for calling a product green, and these can occur in any phase of its life cycle. Commonly, products are designated as environmentally preferable materials because they are:

- Locally harvested or extracted and manufactured
- Sustainably or organically grown and harvested
- Made from **rapidly renewable materials**, those that can naturally be replenished in a short period of time (for LEED, within 10 years)
- Contain recycled content
- Made of biodegradable or compostable material
- Free of toxins
- Long lasting, durable, and reusable
- Made in factories that support human health and workers' rights

For consumers the biggest challenge is identifying what products are truly green. As public interest in sustainability has grown, so has the practice of **greenwashing,** or presenting misinformation to the consumer to portray a product or policy as being more environmentally friendly than it actually is.

- **Specify green materials and equipment.** Give preference to rapidly renewable materials, regional materials, salvaged materials, and those with recycled content. Choose vendors who promote source reduction through reusable or minimal packaging of products. Look for third-party certifications, such as the Forest Stewardship Council, Green Seal, and ENERGY STAR.

- **Specify green custodial products.** Choose sustainable cleaning products and materials that meet Green Seal, Environmental Choice, or EPA standards to protect indoor environmental quality and reduce environmental damage.

WASTE MANAGEMENT

Building construction generates large amounts of solid waste, and waste is generated across the building life cycle as new products arrive and used materials are discarded. This waste may be transported to landfills, incinerated, recycled, or composted. Solid waste disposal contributes directly to greenhouse gas emissions through transportation and, perhaps more significantly, the production of methane—a potent greenhouse gas—in landfills. Incineration of waste produces carbon dioxide as a byproduct. EPA has estimated greenhouse gas emissions from building waste streams and finds that the United States currently recycles approximately 32% of its solid waste—the carbon dioxide equivalent of removing almost

40 million cars from the road. Improving recycling rates to just 35% could result in savings equivalent to more than 5 million metric tons of carbon dioxide.[29]

The intent of LEED credits in this category is to reduce the waste that is hauled to and disposed of in landfills or incineration facilities. During construction or renovation, materials should be recycled or reused whenever possible. During the building's daily operations, recycling, reuse, and reduction programs can curb the amount of material destined for local landfills.

Strategies to reduce waste during construction:

- **Develop a construction waste management policy.** Outline procedures and goals for construction waste diversion. This policy should specify a target diversion rate for the general contractor.

- **Establish a tracking system.** Ensure that the general contractor provides waste hauler reports and captures the full scope of the waste produced. Designate a construction and demolition waste recycling area. Diligent monitoring will ensure that the policy is effective.

29 U.S. Environmental Protection Agency, *Measuring Greenhouse Gas Emissions from Waste* (2010), http://www.epa.gov/climatechange/wycd/waste/measureghg.html

> **Strategies to reduce waste during operations and maintenance:**
>
> - **Develop a solid waste management policy.** Outline procedures and goals for solid waste diversion. This policy should specify a target diversion rate for the facility.
>
> - **Conduct a waste stream audit.** Establish baseline performance for the facility and identify opportunities for increased recycling, education, and waste diversion.
>
> - **Maintain a recycling program.** Provide occupants with easily accessible collectors for recyclables. Label all collectors and list allowable materials. Through signage or meetings, educate occupants about the importance of recycling and reducing waste.
>
> - **Monitor, track, and report.** Use hauler reports or other reliable data to monitor and track the effectiveness of the policy. Track performance goals and provide feedback to the occupants.
>
> - **Compost.** Institute an on-site composting program to turn landscaping debris into mulch. Work with the waste hauler to allow for collection and composting of food and other organic materials.
>
> - **Provide recycling for durable goods.** Institute an annual durable goods drive where e-waste and furniture are collected on site and disposed of properly through donation, reuse, or recycling. Allow occupants to bring e-waste and furniture from home.

LEED IN PRACTICE

LEED for Existing Buildings: Operation & Maintenance encourages building managers to embrace new attitudes and new mindsets and close the life cycle loop by reusing and recycling on-site materials. Understanding the content of a waste stream is the first step to improving the waste diversion rate at a facility.

To comply with LEED requirements, a project team must conduct a waste stream audit for the entire consumables waste stream. The audit results are used to establish a baseline that identifies the

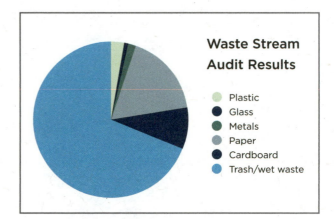

Figure 4.5. Waste Stream Audit Results

amount and percentage of each material in the waste stream. Results from the waste audit can reveal opportunities for increasing recycling and waste diversion and be used to adjust the recycling procedures at the facility.

Assume that a project team has conducted a waste stream audit and tracked 300 pounds of waste, consisting of the following:

	Pounds	Percentage
Trash and wet waste	200	68
Paper	60	20
Cardboard	25	8
Plastic	6	2
Metal	5	1
Glass	4	1

Because 28% of the waste stream is recyclable paper and cardboard paper, the project team should provide recommendations to improve the recycling rate and source reduction of these items. The team should also share the audit results with the building's occupants to encourage their participation in on-site recycling programs.

INDOOR ENVIRONMENTAL QUALITY

Indoor Environmental Quality (IEQ) encompasses the conditions inside a building—air quality, lighting, thermal conditions, ergonomics—and their effects on occupants or residents. Strategies for addressing IEQ include those that protect human health, improve quality of life, and reduce stress and potential injuries. Better indoor environmental quality can enhance the lives of building occupants, increase the resale value of the building, and reduce liability for building owners. Additionally, since the personnel costs of salaries and benefits typically surpass operating costs of an office building, strategies that improve employees' health and productivity over the long run can have a large return on investment. IEQ goals often focus on providing stimulating and comfortable environments for occupants and minimizing the risk of building-related health problems.

To make their buildings places where people feel good and perform well, project teams must balance selection of strategies that promote efficiency and conservation with those that address the needs of the occupants and promote well-being. Ideally, the chosen strategies do both: the solutions that conserve energy, water and materials also contribute to a great indoor experience.

LEED addresses the following issues related to indoor environmental quality:

● Indoor air quality

● Occupants' well-being, comfort, and control

INDOOR AIR QUALITY

The quality of air outdoors has received considerable attention in recent decades, and strategies to reduce smog and other air pollutants are vitally important. However, the air we breathe indoors—where millions of Americans spend most of their day—can be even more polluted. Many common sources generate indoor air contaminants:

● People smoking tobacco inside the building or near building entrances or air uptakes

● Building materials such as paints, coatings, adhesives, sealants, and furniture that may emit **volatile organic compounds (VOCs)**, substances that vaporize at room temperature and can cause health problems

● Combustion processes in HVAC equipment, fireplaces and stoves, and vehicles in garages or near entrances

● Mold resulting from moisture in building materials

● Cleaning materials

● Radon or methane off-gassing from the soil underneath the building

● Pollutants from specific processes used in laboratories, hospitals, and factories

● Pollutants tracked in on occupants' shoes

● Occupants' respiration, which increases carbon dioxide levels and may introduce germs

The best way to prevent indoor pollutants is to eliminate or control them at the sources. The next line of defense is proper ventilation to remove any pollutants that do enter. Both approaches need to be considered at all phases of the building life cycle.

Strategies for improving indoor air quality during construction:

● **Prohibit smoking.** Institute a no-smoking policy in the building and around building entrances, operable windows, and air intakes.

● **Protect air that comes into the building.** Locate air intakes away from likely exhaust sources, such as idling vehicles or smoking areas. Locate smoking areas away from building entrances.

- **Specify low-emitting materials.** Use green materials for both new construction and renovations. Select low-VOC paints, adhesives, sealants, and furniture.

- **Develop and follow a construction indoor air quality management plan.** The plan should include dust control and good housekeeping, protection of pervious materials from moisture, and protection and capping of ducts and mechanical systems.

- **Test for radon or other on-site contaminants.** If present, include a ventilation system to address possible emissions.

- **Design for proper ventilation.** Consider the number of occupants in each space and the activities they will be engaged in. Make sure that the ventilation system, whether natural or mechanical, can provide enough air exchanges. Size the systems appropriately.

- **Use air filters with high MERV ratings. Minimum efficiency reporting value (MERV) Rating** is a measurement scale designed by ASHRAE to rate the effectiveness of air filters. The higher the MERV rating the greater particulates captured by a filter.

- **Protect air quality during construction.** Prevent mold by protecting all materials from moisture exposure. Prevent dust and particulate buildup.

- **Conduct a flush-out.** Before occupancy, flush out indoor airborne contaminants by thoroughly exhausting old air and replacing it with fresh, outdoor air.

- **Install entryway grates.** Use permanently installed, cleanable grates or mats to remove pollutants carried by people's shoes.

Indoor air quality must be maintained throughout the life of a building to protect occupants on an ongoing basis.

Strategies for improving indoor air quality during operations and maintenance:

- **Ensure adequate ventilation.** Operate ventilation systems to supply ample outside air to the occupants. Follow the most recent industry standards, such as ASHRAE Standard 62, Ventilation for Acceptable Indoor Air Quality.

- **Monitor outdoor airflow.** Use an outdoor airflow measurement device that can measure and control the minimum outdoor airflow rate.

- **Monitor carbon dioxide.** Use monitors and integrate them with a ventilation system that regulates the supply of air based on occupants' demand. With demand-controlled ventilation, air flow is automatically increased if concentrations exceed a setpoint.

- **Prohibit smoking.** Enforce a no-smoking policy in the building and around building entrances, operable windows, and air intakes. Communicate the policy to building occupants through building signage and tenant meetings.

- **Calibrate sensors.** Perform routine preventive maintenance, such as calibrating sensors and monitors, to ensure that accurate data are used to modulate systems.

- **Develop and implement a green cleaning policy.** To minimize the introduction of contaminants, outline procedures and goals for the custodial program at the facility. This policy should specify standards for selecting cleaning products and technologies, such as Green Seal standards, California Code of Regulations, and certification of cleaning equipment from the Carpet and Rug Institute.
- **Conduct custodial effectiveness assessment.** Identify opportunities for improving building cleanliness and reducing occupants' exposure of potentially harmful biological and particulate contaminants.
- **Employ permanent entryway systems.** Place grilles, grates, or mats at least 10 feet long at all major entrances help to reduce the dust, dirt, and contaminants brought into the facility. Develop cleaning procedures to properly maintain entryway systems.
- **Use integrated pest management.** A coordinated program of nonchemical strategies, such as monitoring and baiting, will reduce the need for pesticides and other potentially toxic contaminants.

OCCUPANTS' WELL-BEING, COMFORT, AND CONTROL

To be healthy, happy, and productive in the building, occupants need to feel comfortable and in control of their environment. This includes thermal comfort, lighting and views, acoustics, and ergonomics. Feeling too hot or too cold, having insufficient lighting or being unable to look out a window, dealing with too much noise or having an uncomfortable work station can all cause stress and reduce quality of life. Because people's needs vary and even the same individual may have different needs and preferences at different times, the ability to control the indoor environment is a critical component of occupants' comfort and satisfaction.

Thermal comfort includes more than just temperature; it also includes humidity and air movement. An area may be the right temperature, but if the air is stagnant or if air ducts blow directly on their work stations, people will feel uncomfortable. An operable window may make office workers more comfortable than a sealed environment maintained at ideal temperatures simply because it gives them some control over their environment.

Daylit Classroom
photo credit: Josh Partee 2009

Lighting levels and views to the outdoors are other important aspects of the indoor experience. Providing enough lighting for particular tasks is critical to protect occupants' eyesight over time. Studies by the Heschong Mahone Group have demonstrated that providing daylighting in classrooms can improve student scores by 7% to 18%.[30] They also found improvements in office workers' productivity. In addition to admitting daylight, windows that let people focus their eyes across a longer distance and see the outdoors may play a role in occupants' comfort. Of course, too much light can interfere with some tasks, and direct sunlight or glare can create discomfort as well. Good lighting design considers the tasks to be done in a space, the orientation of the building, the layout of the room, the type of glass and configuration of the windows, even the type of furnishings and colors of surfaces.

Appropriately sized and located windows can dramatically increase the amount of daylight introduced into a space; clerestory windows, light shelves, and reflective paint and materials bounce and diffuse the natural light. In office buildings, locating private offices toward the building core and siting cubicles at the perimeter brings daylight into a large area. Low cubicle partitions allow daylight to travel to the core spaces while permitting views of the outdoors. Adjustable window shades give occupants control over excessive brightness and glare.

Daylight can also decrease the need for artificial lighting. Daylight controls help in dimming or turning off electrical lights entirely when daylight is sufficient. These controls should be zoned so that the spaces near the windows, with lots of natural light, have dimmed artificial lighting, and the spaces farther away from the perimeter, with less natural light, have higher levels of artificial light.

When designing buildings, consider energy conservation and IEQ together. It is easy to view these considerations as contradictory. However, a systems-based, integrated approach can identify solutions that contribute to both goals. For example, natural daylighting and ventilation can not only save energy but also improve occupants' experience. Furthermore, once the building design team members understand who the occupants are, what they will be doing, and how they will be doing it, they can create environments tailored to those needs while providing sufficient control and flexibility.

IEQ systems must be evaluated and adjusted once the building is occupied. Installing sensors to monitor conditions and conducting occupant surveys are important parts of green building operations.

30 Heshong Mahone Group, *Windows and Offices: A Study of Office Worker Performance and the Indoor Environment* (CEC PIER, 2003), http://www.h-m-g.com/projects/daylighting/summaries%20on%20daylighting.htm.

- **Use daylighting.** Design the building to provide ample access to natural daylight and views for the occupants. Optimize access to views by using low partitions and vision panels.

- **Install operable windows.** If possible, provide windows that can be opened to the outside. To save energy, sensors may be included to inform the HVAC system to shut down if a window is open.

- **Give occupants temperature and ventilation control.** In mechanically ventilated buildings, provide thermostats that allow occupants to control the temperature in their immediate environment. Provide adjustable air diffusers that allow occupants to adjust the air flow as well.

- **Give occupants lighting control.** Provide adjustable lighting controls so that occupants can match lighting levels to their tasks. These may be designed in combination with daylight and occupancy sensors to conserve energy.

- **Conduct occupant surveys.** Use valid survey protocols to assess occupants' satisfaction with the indoor environment. Evaluate results to identify areas of dissatisfaction and prepare a corrective action plan to make the necessary operational changes.

- **Provide ergonomic furniture.** Include furniture that is adjustable to prevent repetitive stress injuries.

- **Include appropriate acoustic design.** Use soft surfaces and other strategies to ensure that sound levels remain comfortable for the activity level of the space.

LEED IN PRACTICE

LEED for Existing Buildings: Operations & Maintenance encourages facilities managers to assess occupants' comfort levels while at work. Through a confidential survey, occupants can rate the heating and air-conditioning, acoustics, air quality, lighting levels, cleanliness, and other aspects of their work spaces. Facilities managers evaluate the responses to determine any areas of dissatisfaction, then develop a corrective action plan to address problems and improve occupants' comfort. Figure 4.6 illustrates an example developed at the University of California-Berkeley in which occupant comfort survey results are shown using a 7 point scale.

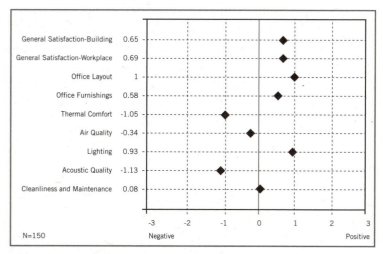

Figure 4.6. Sample Occupant Survey
(Source: LEED Reference Guide for Green Building Operations and Maintenance, Washington, DC, 2009.)

INNOVATION
IN DESIGN AND
OPERATIONS

Through the Innovation category, LEED encourages and recognizes project team efforts to create additional environmental benefits beyond those already achieved through other rating system categories. Innovative strategies expand the breadth of green building practice by incorporating cutting-edge techniques, processes and products into the development of a project. Ideally, innovation is a byproduct of the green building process discussed in this guide. The integrated and iterative processes required to achieve the environmental benefits addressed by LEED encourage new methods and standards, while advancing the practice of green building.

Projects incorporating these strategies and achieving exemplary levels of performance are rewarded with innovation credits. Three basic practices can lead projects to achieve innovation credits for a category not specifically addressed by LEED are as follows:

1. The project must demonstrate quantitative performance improvements for environmental benefit by establishing a baseline of standard performance and comparing that with the final design that includes innovative strategies.

2. The process or specification must be comprehensive. In other words, a team that is considering applying for an Innovation credit must demonstrate that the program applies to the entire project being certified under LEED, as opposed to only addressing a limited portion of the project.

3. The concept must be replicable and applicable to other project. It must be significantly better than standard sustainable design principles and practices.

Strategies and practices rewarded as innovative today may become credits in future rating systems. In fact, as LEED continues to evolve and today's innovation become tomorrow's standard, strategies that may have earned Innovation credit in the past may not necessary earn recognition today.

Examples of innovative strategies include:

Developing a comprehensive educational outreach program that encourages the advancement of the community, occupant, resident or other stakeholders' knowledge of the characteristics of green building and how best to achieve and take advantage of them.

- Evaluating a large quantity of products being used in the project and demonstrate that they provide significant performance advantages or environmental benefits, based on an acceptable life-cycle assessment.

- Creating, implementing and maintaining a program for occupants or other stakeholders to divert a significant amount of waste generated from outside sources to appropriate recycling locations.

Project Case Study

©The Kubala Washatko Architects, Inc./Mark F. Heffron

THE ALDO LEOPOLD LEGACY CENTER

The Aldo Leopold Legacy Center, near Baraboo, Wisconsin, was the first building recognized by USGBC as carbon neutral— an exceptional achievement that helped the project earn all 5 points under the Innovation in Design credit category. The project team prepared a greenhouse gas emissions budget based on the requirements of the World Resources Institute Greenhouse Gas Protocol. Conservatively accounting for carbon generation and sequestration in metric tons of CO_2 equivalent (a measure of greenhouse gas emissions that combines multiple heat-trapping gases, such carbon dioxide, methane, and nitrous oxide), the activities of the center will result in the net *reduction* of CO_2 emissions each year.

Projected annual greenhouse gas emissions from Aldo Leopold Legacy Center

	CO_2 equivalent per year (metric tons)	
Total emissions		13.42
Offset from renewable energy	−6.24	
Onsite forest sequestration	−8.75	
Total emissions reduction		−14.99
Net balance of emissions		−1.57

More information about the Aldo Leopold Legacy Center is available at http://www.aldoleopold.org/legacycenter/carbonne utral.html.

SECTION 5

U.S. GREEN BUILDING COUNCIL AND ITS PROGRAMS

The U.S. Green Building Council (USGBC), a nonprofit organization, is a coalition of leaders from every sector of the building industry working to promote environmentally responsible, profitable, and healthful places to live, learn, and work. USGBC members represent more than 15,000 organizations and 55 professions, ranging from real estate professionals and building owners and managers to lawyers, architects, engineers, and contractors.

HISTORY OF USGBC

In the early 1990s committed visionaries in the architecture, engineering, and construction industries and related areas came together to breathe new life into the green technology industry. This field had quietly simmered since the 1970s energy crisis, which sparked initial interest and investment in efficiency and renewable energy sources. As the 20th century drew to a close, these leaders took fresh inspiration from the writings of pioneers such as Buckminster Fuller, as well as from new initiatives like the Greening of the White House. Their common interests made unlikely allies of typically disparate areas of the building industry, bringing together environmentally conscious designers and business leaders. USGBC provided a platform for these groups to have thoughtful conversations, consolidate ideas about green construction, and establish a framework to support enhanced building performance.

The organization's founding members quickly realized that the sustainable building industry needed a system to define and measure green buildings. USGBC began to research existing building metrics and rating systems, and less than a year after formation, the members acted on the initial findings by establishing a committee to focus on this topic. The composition of the committee was diverse: architects, real estate agents, a building owner, a lawyer, an environmentalist, and industry representatives. This cross section of people and professions added a richness and depth both to the process and to the outcome. The initial process undertaken by this group embodied the community-based governance process that remains at the core of USGBC's work. The committee laid the foundation for the development of the LEED rating system and ideas that would quickly become mainstream.

> **USGBC'S MISSION: TO TRANSFORM THE WAY BUILDINGS AND COMMUNITIES ARE DESIGNED, BUILT, AND OPERATED, ENABLING AN ENVIRONMENTALLY AND SOCIALLY RESPONSIBLE, HEALTHY, AND PROSPEROUS ENVIRONMENT THAT IMPROVES THE QUALITY OF LIFE.**

USGBC TODAY

Today USGBC continues to pursue its vision of buildings and communities that will regenerate and sustain the health and vitality of all life within a generation. Based in Washington, D.C., USGBC's dedicated staff and expansive community of volunteers develop new products, services, and programs each year in several major areas of work.

ADVOCACY

USGBC provides policymakers and community leaders with the tools, strategies, and resources they need to take leadership positions, foster innovation, and inspire action. From national advocacy programs promoting green schools to policy engagement with decision makers in the White House and the U.S. Congress, as well as state houses and city halls across the country, USGBC is accelerating the uptake of policies and initiatives that enable and encourage market transformation toward a sustainable built environment.

COMMUNITY

The USGBC community comprises member organizations that participate in forums, exchanges, and regular communication. Additionally, there are regional USGBC chapters and affiliates across the nation. This network of industry leaders provides green building resources, education, and opportunities for green building professionals, both those established in the field and those who are new to the practice, to stay connected in their communities.

EDUCATION

USGBC provides high-quality educational programs and materials on green design, construction, and operations for professionals from all sectors of the building industry. The focus is on developing practical knowledge, exploring new business opportunities, and learning how to create more healthful, productive, and efficient places to live and work. USGBC's diverse delivery formats, including instructor-led training, webinars, online courses, and publications, make learning about green building accessible to all.

GREENBUILD INTERNATIONAL CONFERENCE AND EXPO

Greenbuild is the world's largest conference and exposition dedicated to green building. Launched in 2002, it has become an important annual event for the green building industry. Each year, tens of thousands of professionals convene to take part in educational sessions, tour green buildings, and view exhibits of green products and technologies.

LEED® GREEN BUILDING RATING SYSTEM™

USGBC's Leadership in Energy and Environmental Design (LEED) program encourages and accelerates adoption of sustainable building and community development practices through the creation and implementation of a green building benchmark that is voluntary, consensus based, and market driven. The technical basis of LEED is existing and emerging standards, tools, and performance criteria. LEED seeks a balance between requiring today's best practices and encouraging innovative strategies; its rating systems are a challenging yet achievable set of building and neighborhood benchmarks that define green building around the world.

LEADERSHIP IN ENERGY AND ENVIRONMENTAL DESIGN

LEED is a third-party green building certification program and the nationally accepted benchmark for the design, construction, and operation of high-performance green buildings and neighborhoods. The rating systems give building owners and operators the tools they need to have an immediate and measurable effect on their buildings' performance. By promoting a whole-building approach to sustainability, LEED recognizes performance in location and planning, sustainable site development, water savings, energy efficiency, materials selection, indoor environmental quality, innovative strategies, and attention to priority regional issues. Additionally, LEED addresses all building types through different rating systems and rating system adaptations.

LEED RATING SYSTEMS

Comprehensive and flexible, LEED is applicable to buildings at any stage in their life cycles. New construction, the ongoing operations and maintenance of an existing building, and a significant tenant retrofit to a commercial building are all addressed by LEED rating systems.

The rating systems and their companion reference guides help teams make the right green building decisions for their projects through an integrated process, ensuring that building systems work together effectively. Through a consensus-based process, the rating systems are continually evaluated and regularly updated to respond to new technologies and policies and to changes in the built environment. In this way, as yesterday's innovation becomes today's standard of practice, USGBC and LEED continue to push forward market transformation.

The following project types and scopes are addressed by LEED rating systems:

- LEED for New Construction and Major Renovations
- LEED for Core and Shell
- LEED for Commercial Interiors
- LEED for Schools
- LEED for Healthcare
- LEED for Retail
- LEED for Existing Buildings: Operations and Maintenance
- LEED for Homes
- LEED for Neighborhood Development

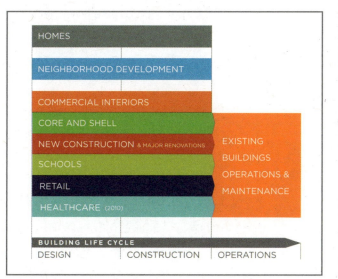

Figure 5.1. LEED Rating Systems

RATING SYSTEM STRUCTURE

The LEED rating systems consist of prerequisites and credits. Prerequisites are required elements or green building strategies that must be included in any LEED-certified project. Credits are optional elements—strategies that projects can elect to pursue to gain points toward LEED certification.

Achieving LEED certification requires satisfying all prerequisites and earning a minimum number of credits. Each LEED rating system corresponds to a LEED reference guide that explains credit criteria, describes the benefits of complying with the credit, and suggests approaches to achieving credit compliance.

Although the organization of prerequisites and credits varies slightly depending on the building type and associated rating system, LEED is generally organized by the following broad concepts:

- **Sustainable Sites.** Choosing a building's site and managing that site during construction are important considerations for a project's sustainability. LEED credits addressing sustainable sites discourage development of previously undeveloped land and damage to ecosystems and waterways; they encourage regionally appropriate landscaping, smart transportation choices, control of stormwater runoff, and reduced erosion, light pollution, heat island effect, and construction-related pollution. LEED also emphasizes location and transportation issues by rewarding development that preserves environmentally sensitive places

and takes advantage of existing infrastructure, community resources, and transit. It encourages access to open space for walking, physical activity, and time spent outdoors.

- **Water.** Buildings are major users of our potable water supply. The goal of credits addressing water efficiency is to encourage smarter use of water, inside and out. Water reduction is typically achieved through more efficient appliances, fixtures, and fittings inside and water-wise landscaping outside.

- **Energy.** LEED encourages a wide variety of strategies to address energy consumption, including commissioning; energy use monitoring; efficient design and construction; efficient appliances, systems, and lighting; and the use of renewable and clean sources of energy, generated on-site or off-site.

- **Materials and Resources.** During both construction and operations, buildings generate large amounts of waste and use tremendous volumes of materials and resources. These credits encourage the selection of sustainably grown, harvested, produced, and transported products and materials. They promote the reduction of waste as well as reuse and recycling, and they take into account the reduction of waste at a product's source.

- **Indoor Environmental Quality.** The average American spends about 90% of the day indoors, where pollutant concentrations may be two to 100 times higher than outdoor levels. Thus indoor air quality can be significantly worse than outside. LEED credits promote strategies that can improve indoor air, provide access to natural daylight and views, and improve acoustics.

- **Awareness and Education.** A building's occupants need to understand what makes their building green and have the tools to make the most of its features. The LEED for Homes rating system has a separate category to emphasize the role homebuilders and real estate professionals play in interpreting these systems and features for homeowners. In rating systems geared toward commercial buildings, awareness and education are addressed under Innovation.

- **Innovation.** LEED promotes innovation in design and operations by offering bonus points for improving a building's performance well beyond what is required by the credits or for incorporating green building ideas that are not specifically addressed elsewhere in the rating system. This credit category also rewards the inclusion of a LEED Accredited Professional on the project team. Additionally, teams may earn credit in this category for an education plan that shares green building information with occupants and the public.

- **Regional Priority.** USGBC's regional councils, chapters, and affiliates have identified the environmental concerns that are most important for every region of the country, and six LEED credits that address those local priorities have been selected for each region. A project team that earns a regional priority credit earns one bonus point in addition to any points awarded for that credit. Up to four extra points can be earned in this way.

LEED for Neighborhood Development diverges significantly from other rating systems and is organized around three main categories, focusing on where, what, and how to build green at a community scale.

- **Smart Location and Linkages.** This section of the rating system provides guidance on where the project is built, encouraging the selection of sites with existing services and transit.

- **Neighborhood Pattern and Design.** Neighborhoods should be compact, complete, connected, and convivial. The intent of credits in this category is to create environments that are walkable, vibrant with mixed-use establishments, and connected to the larger community.
- **Green Infrastructure and Buildings.** This category focuses on measures that can reduce the environmental harms associated with the construction and operation of buildings and infrastructure within neighborhoods, with a goal of not just reducing the environmental consequences, but also enhancing the natural environment.

Additionally, LEED emphasizes the critical role of the integrated process and ongoing **performance monitoring** across all phases and project types.

LEED rating systems generally have 100 base points plus six Innovation points and four Regional Priority points, for a total of 110 points. The level of certification for commercial projects is determined according to the following scale:

- Certified, 40–49 points
- Silver, 50–59 points
- Gold, 60–79 points
- Platinum, 80+ points

LEED for Homes certification levels vary slightly because the rating system is based on a 125-point scale, plus 11 innovation points.

RATING SYSTEM DEVELOPMENT AND EVOLUTION

Since its launch in 2000, LEED has been evolving to address new markets and building types, advances in practice and technology, and greater understanding of the environmental and human health impacts of the built environment. These ongoing improvements to LEED are based on principles of transparency, openness, and inclusiveness involving volunteer committees and working groups, as well as USGBC staff, and approval by a membership-wide vote.

LEED is updated through a regular development cycle for revisions to the rating system. There are three basic types of LEED development:

- **Implementation and maintenance of current version.** LEED rating systems are continually improved through the correction and clarification of credit language. These updates are published as quarterly addenda and include LEED interpretations (see below).
- **LEED rating system adaptations.** Credit adaptations address both specific space types and international projects, meeting the needs of projects that would otherwise be unable to participate in LEED. Currently, four adaptations are available: LEED for Schools and LEED for Healthcare, both derivatives of LEED for New Construction, and LEED for Retail, adapted for both the LEED for New Construction and Commercial Interiors rating system.

- **Next version of LEED.** A periodic evaluation and revision process leads to comprehensive improvement of the rating systems. This phase includes multiple avenues for stakeholder input and final approval by USGBC members. The ideas generated during the development of next-version LEED credits are often pilot-tested by LEED project teams prior to ballot.

Additionally, the **LEED Pilot Credit Library** plays an important role in the evolution of LEED. Pilot credits are tested across all rating system types and credit categories and include credits proposed for the next version of LEED. Project teams may attempt any of these pilot credits under the Innovation categories and earn points by providing USGBC with feedback on the credits' efficacy and achievability. USGBC collects and integrates this feedback to refine the pilot credits, and worthwhile credits are then added to the balloted LEED rating system.

CREDIT WEIGHTINGS

The LEED rating system has always been implicitly weighted by virtue of the different point values assigned to each credit and category. These weightings continue to evolve with the rating system as market conditions, user requirements, scientific understanding and public policy change. The weightings ensure that LEED assigns higher point values to the most important credits and categories. Thus a given credit's point value reflects its potential both to mitigate the environmental harms of a building and to promote beneficial effects.

Deciding which environmental impacts LEED should address led to the initial credit weightings. LEED 2009 used the U.S. Environmental Protection Agency's TRACI environmental impact categories. TRACI is a computer software tool developed by EPA to assist with impact assessment for life cycle assessment, industrial ecology, process design, and pollution prevention. Layered on top of the TRACI environmental impact categories are weightings devised by the National Institute of Standards and Technology that compare the categories and assign

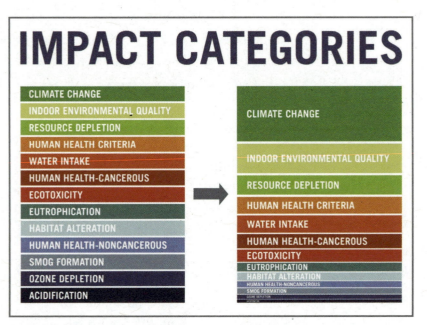

Figure 5.2. TRACI Environmental Impact Categories and LEED

a relative importance to each. The result is a weighted average that combines building effects and the relative value of the impact categories. Overall, the credit weights emphasize energy efficiency, renewable energy, reduced transportation demand, and water conservation, based on their direct contribution to reducing high-priority problems, particularly greenhouse gas emissions.

As the rating systems continue to evolve, credit weightings are increasingly underscoring the capacity of the built environment to have positive effects on natural and human environments.

CARBON OVERLAY

USGBC is a signatory to the **Wingspread Principles on a U.S. Response to Global Warming,** a set of propositions signed by individuals and organizations declaring their commitment to addressing climate change. These principles recognize that the critical challenge of climate change requires changes in our economy, policies, and behaviors. The Wingspread Principles call for urgent and effective action to reduce greenhouse gas emissions by 60% to 80% below 1990 levels by midcentury. As a signatory and in accord with USGBC's mission and vision, LEED offers a powerful tool for lowering greenhouse gas emissions associated with the built environment.

Some green buildings can achieve greater greenhouse gas reductions than others, whether because of the specific strategies used to achieve LEED certification or because of the circumstances of the project, such as its location or source of purchased electricity. Accordingly, USGBC now identifies and prioritizes LEED credits based on their relative value for greenhouse gas emissions reduction. This "carbon overlay" is a quantitative index of the relative importance of individual credits.

The score for each LEED credit is estimated based on the carbon footprint for a typical LEED building. A building's carbon footprint is the total greenhouse gas emissions associated with its construction and operation, including

- Energy used by building systems
- Building-related transportation
- Embodied emissions of water (electricity used to extract, convey, treat, and deliver water)
- Embodied emissions of solid waste (life cycle emissions associated with solid waste)
- Embodied emissions of materials (emissions associated with the manufacture and transport of materials)

The carbon footprint for constructing and operating a typical 135,000-square-foot office building is represented in Figure 5.2, which illustrates the sources of emissions of the annual total of 4,700 metric tons of carbon dioxide equivalent. This distribution can be used to prioritize credits based on their potential to reduce greenhouse gas emissions. The credits addressing the most important emissions sources receive the highest scores in the carbon overlay.

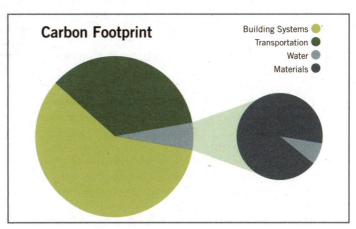

Figure 5.3. Carbon Footprint Distribution of Typical Office Building

GREEN BUILDING CODES, STANDARDS AND RATING SYSTEMS

A growing number of state and local governments are analyzing and revising their building codes to better align with their sustainability goals and green building programs. Even where codes are determined at the state level, many local governments are finding that regulatory minimums for the private sector may need upgrades and more comprehensive enforcement strategies to improve public health, safety, and environmental quality.

A landmark national effort to codify green building practices into adoptable, adaptable and enforceable green building codes has produced regulatory documents that are now available as an overlay to more traditional building codes. The International Green Construction Code (IGCC), including ASHRAE Standard 189.1 as an alternate path to compliance, is a widely supported and first-of-its-kind regulatory framework that recognizes an entire set of risks not otherwise addressed in the codes.

Both distinct and complementary to green building rating systems such as LEED, green building codes are redefining the fundamental protections that are the basis of smart public policy. And, as the floor is raised through the codes, so too is the ceiling raised through beyond-code rating systems like LEED that continue to pave the way, constantly raising the bar for leadership.

GREEN BUILDING CERTIFICATION INSTITUTE

In 2008, the Green Building Certification Institute (GBCI) was established as an independent, third-party incorporated entity with the support of USGBC. GBCI administers project certifications and professional credentials and certificates for the LEED rating systems. Whereas USGBC handles the development of the LEED rating systems and offers LEED-based education and research programs, GBCI administers the LEED Professional Accreditation program independently, to allow for balanced, objective management of the credential. Additionally, GBCI performs reviews and awards LEED certification to projects.

PROJECT CERTIFICATION

LEED certification provides independent, third-party verification that a building project meets the highest green building and performance measures. Early in the development of a project, the integrated project team needs to determine the project's goals, the level of certification to pursue, and the credits that will help them achieve it. The certification steps generally proceed as follows.

Project Registration

The LEED process begins with registration. The project team submits a registration form and a fee to GBCI. It is helpful if the **project administrator**—the team member who registers

the project—has previous green building and LEED project experience; ideally, he or she is a LEED Accredited Professional. Once registered, the team receives information, tools, and communications that will help guide the certification process. All project activity, including registration and credit compliance documentation, is completed in **LEED Online**, a data collection portal through which the team uploads information about the project. This site provides credit templates to be completed and signed by a specified member of the team.

Application Preparation

Each LEED credit and prerequisite has documentation requirements that must be completed as part of the application process. The project team selects the credits it has chosen to pursue and assigns responsibility for each credit to a team member. When the necessary documentation, including required information and calculations, has been assembled, the project team uploads the materials to LEED Online.

Submission

When the team is ready for GBCI to review its application, the project administrator submits the appropriate fee, which is based on project square footage, and documentation. Submission of documentation for review can be done in either one or two stages. The team can wait to submit documentation until the building project is complete, or the team can seek review of its design-related prerequisites and credits before completion, and then apply for construction-related credits after the project is finished.

Application Review

Whether the design and construction credits are submitted together or separately, each credit undergoes one preliminary review. The GBCI reviewer may request additional information or clarification. The team then submits final documentation. After the final review, a team may appeal any adverse decisions on individual credits to GBCI for an additional fee.

Certification

Certification is the final step in the LEED review process. Once the final application review is complete, the project team can either accept or appeal the final decision. LEED-certified projects receive formal certificates of recognition, information on how to order plaques, and tips for marketing the achievement. Projects may be included in USGBC's online LEED Project Directory of registered and certified projects.

Variants

Two LEED rating systems follow slightly different processes, reflecting differences in planning and development.

LEED for Homes involves a multistep review, based on the phases in the design and construction of new homes, with participation by project team members, including a LEED for Homes provider, the homebuilder, a green rater, and a home energy rater. The verification process has five steps:

- Step 1. Early planning
- Step 2. Design
- Step 3. Build
- Step 4. Verification and certification

LEED for Neighborhood Development follows the typical neighborhood development process—which can take years or even decades—from the earliest steps of project entitlement (acquiring the necessary permits) to the completion of the project. This process is segmented into three stages for review:

- Stage 1. Review prior to completion of the entitlement (permitting) process
- Stage 2. Certification of an approved development plan
- Stage 3. Review of a completed neighborhood development

USGBC provides comprehensive information about the certification process within the LEED section of its website.

PROJECT CREDIT INTERPRETATION RULINGS AND LEED INTERPRETATIONS

The LEED rating systems are intended to be flexible, voluntary tools to improve the performance of buildings and promote market transformation. At times a project team may want clarification, further guidance, or additional ways to comply with the rating system's requirements. Project teams therefore have several options in engaging with USGBC and GBCI during the certification submittal process.

Project credit interpretation rulings (Project CIRs), administered by GBCI, allow teams to obtain technical guidance on how LEED requirements pertain to their projects. Project CIRs do not guarantee credit award; the project applicant must still demonstrate and document achievement during the LEED certification application process. A team seeking a Project CIR uses LEED Online's formal inquiries tab. Guidelines for requesting a Project CIR are available on GBCI's website. The ruling remains confidential and generally applies only to the one project.

Project teams can contribute to the evolution of LEED in a significant way through **LEED Interpretations** which can generate precedent-setting rulings. This process begins when a project team submits an inquiry to GBCI and chooses to opt in to the LEED Interpretations process. This request is first assessed by a certification reviewer. The initial ruling may undergo a subsequent, consensus-based review by the LEED committee and staff, whose ruling can

be applied to multiple buildings, LEED rating systems, and programs—by the same project team and by others. The database of LEED Interpretations and Addenda is online. Additional guidance on the LEED interpretations process can be found on USGBC's website.

LEED PROFESSIONAL CREDENTIALS

GBCI also manages all aspects of the LEED Professional Accreditation program, including exam development, registration, and delivery. LEED professionals demonstrate current knowledge of green building technologies, best practices, and the rapidly evolving LEED rating systems. The credentials differentiate professionals in a growing and competitive industry, allow for varied levels of specialization, and give employers, policy makers, and other stakeholders an indication of individuals' level of competence. Credential holders must demonstrate their knowledge by first qualifying and then sitting for an exam. Eligibility requirements vary based on the level of accreditation a candidate seeks. Accreditation is available at three levels:

- **LEED Green Associate** validates basic understanding of green building and the professional field, as gained through experience in sustainability and green building or related educational experience
- **LEED Accredited Professional** demonstrates a deep familiarity with the LEED rating systems developed through active participation in and contribution to a LEED-registered project
- **LEED Fellow** distinguishes professional leadership, contribution to the standards of practice and body of knowledge, and continual improvement in the field

Additionally, GBCI administers LEED Professional Certificates™ to certify the skills and knowledge of LEED implementation required to provide verification services on LEED projects. Credentialing as a LEED for Homes Green Rater is available now; certificates for LEED Project Reviewer are forthcoming.

To keep their credentials current, professionals must meet continuing education requirements that help them grow their knowledge base, stay current with best practices, and demonstrate that their expertise is meaningful in a rapidly transforming marketplace. GBCI oversees the development and implementation of a credential maintenance program for LEED Accredited Professionals. Establishing continuing education requirements for LEED Accredited Professionals ensures that the credential continues to distinguish those building professionals who have a thorough understanding of green building principles and practices plus the skills to steward the LEED certification process.

CONCLUSION

Equipped with the fundamental knowledge about green building and LEED core concepts, let's return to the building described at the beginning of this guide by USGBC CEO Rick Fedrizzi. Imagine that large oak table in the LEED Platinum commercial office space and take a seat.

As part of your company's finance team, perhaps you are working closely with the property manager to finalize details on a green lease agreement for this space. Or soon after the lease is signed and your company has moved in, you are part of the operations staff sitting around the same table, discussing the steps needed to implement your company's new green cleaning guidelines.

Now, imagine that the oak table is in your LEED Platinum home. You are enjoying a meal with your kids, home from college for summer vacation. As you pass dishes around the table, filled with fresh food from the supermarket you can walk to, your daughter applauds you for taking her advice and pursuing LEED certification for your home.

Next, imagine that you are that son or daughter. You are back on campus, sitting at your own oak table, in your school's LEED Platinum student activities center. You share your experiences volunteering at the most recent Greenbuild Conference with your peers and professors. Inspired by your stories, the team decides to plan an on-campus sustainability conference in the months ahead.

Finally, imagine that you are sitting at that oak table, back in the LEED Platinum commercial office space, flooded by natural springtime light. You work closely with your colleagues, following up on plans laid out in a recent project charrette. Looking up, you take pride in what you can contribute as a LEED Green Associate or LEED Accredited Professional.

Every day, we pass into and out of these buildings, often without giving them much notice. There is tremendous opportunity in our homes, our offices, our schools, our hospitals, our places of worship, and our neighborhoods. As we work together to transform the

built environment, we can find solutions to climate change, water and resource shortages, unemployment and economic distress. We can tackle tough issues like traffic congestion and respiratory illnesses. With the knowledge you now carry, you are not only equipped to take part in the conversations about these challenges—you are prepared to be a leader, find solutions, and transform the built environment.

APPENDICES

APPENDIX A: USGBC & GBCI Resources

LEED Green Associate Candidate Handbook
http://www.gbci.org/Libraries/Candidate_Handbooks/LEED_Green_Associate_Candidate_Handbook.sflb.ashx

USGBC Education Resources
http://www.usgbc.org/education

U.S. Green Building Council Publications, Courses, and Related Products
http://www.usgbc.org/store

Green Building Codes Overview
http://www.usgbc.org/ShowFile.aspx?DocumentID=7403

Information on USGBC and its Interaction with Government
http://www.usgbc.org/government

LEED for Neighborhood Development
http://www.usgbc.org/nd

LEED 2009 for New Construction and Major Renovations Rating System (USGBC, 2009)
http://www.gbci.org/Libraries/Credential_Exam_References/LEED-2009-for-New-Construction-Rating-System.sflb.ashx

LEED for Existing Buildings: Operations & Maintenance Reference Guide, Introduction (USGBC, 2009)
http://www.gbci.org/Libraries/Credential_Exam_References/LEED-for-Operations-and-Maintenance-Reference-Guide-Introduction.sflb.ashx

LEED for Existing Buildings: Operations & Maintenance Reference Guide, Glossary (USGBC, 2008)
http://www.gbci.org/Libraries/Credential_Exam_References/LEED-for-Operations-and-Maintenance-Reference-Guide-Glossary.sflb.ashx

LEED for Homes Rating System (USGBC, 2008)
http://www.gbci.org/Libraries/Credential_Exam_References/LEED-for-Homes-Rating-System.sflb.ashx

LEED 2009 for Operations & Maintenance Rating System (USGBC, 2009)

http://www.gbci.org/Libraries/Credential_Exam_References/LEED-EBOM-Rating-System.sflb.ashx

LEED 2009 Minimum Program Requirements (USGBC, 2009)

http://www.usgbc.org/ShowFile.aspx?DocumentID=6715

Cost of Green Revisited, by Davis Langdon (2007)

http://www.gbci.org/Libraries/Credential_Exam_References/Cost-of-Green-Revisited.sflb.ashx

Sustainable Building Technical Manual: Part II, by Anthony Bernheim and William Reed (1996)

http://www.gbci.org/Libraries/Credential_Exam_References/Sustainable-Building-Technical-Manual-Part-II.sflb.ashx

The Treatment by LEED® of the Environmental Impact of HVAC Refrigerants (LEED Technical and Scientific Advisory Committee, 2004)

http://www.gbci.org/Libraries/Credential_Exam_References/The-Treatment-by-LEED-of-the-Environmental-Impact-of-HVAC-Refrigerants.sflb.ashx

Credit Interpretation Rulings (GBCI)

http://www.gbci.org/Certification/Resources/cirs.aspx

Guidance on Innovation in Design (ID) Credits (USGBC, 2004)

http://www.gbci.org/Libraries/Credential_Exam_References/Guidance-on-Innovation-and-Design-Credits.sflb.ashx

APPENDIX B: Case Study Information

PRAIRIE CROSSING STATION VILLAGE

Year Completed: 2010
Location: Grayslake, Illinois
LEED Certification Level: Certified
Rating System: LEED ND - Pilot
Organization Website: http://www.prairiecrossing.com/

GAIA NAPA VALLEY HOTEL CANYON VALLEY

Year Completed: 2006
Location: American Canyon, California
LEED Certification Level: Gold
Rating System: LEED NC Version 2.1
Organization Website: http://www.gaianapavalleyhotel.com/

KENYON HOUSE

Year Completed: 2008
Location: Seattle, Washington
LEED Certification Level: Platinum
Rating System: LEED for Homes Version 1.0
Organization Website: http://www.buildingchanges.org/news-room/heads-up/132-welcome-home-to-kenyon-house
USGBC Case Study Website: http://www.usgbc.org/ShowFile.aspx?DocumentID=8825

CHARTWELL SCHOOL

Year Completed: 2006
Location: Seaside, California
LEED Certification Level: Platinum
Rating System: LEED NC Version 2.1
Organization Website: http://www.chartwell.org/
USGBC Case Study Website: http://www.usgbc.org/ShowFile.aspx?DocumentID=8824

ONE AND TWO POTOMAC YARD

Year Completed: 2006
Location: Arlington, Virginia
LEED Certification Level: Gold
Rating System: LEED NC Version 2.1 and LEED EB Version 2.0
Organization Website: http://www.potomacyard.net/
USGBC Case Study Website: http://www.usgbc.org/ShowFile.aspx?DocumentID=8827

GLOSSARY

acid rain the precipitation of dilute solutions of strong mineral acids, formed by the mixing in the atmosphere of various industrial pollutants (primarily sulfur dioxide and nitrogen oxides) with naturally occurring oxygen and water vapor.

adapted plants non native, introduced plants that reliably grow well in a given habitat with minimal winter protection, pest control, fertilization, or irrigation once their root systems are established. Adapted plants are considered low maintenance and not invasive.

adaptive reuse designing and building a structure in a way that makes it suitable for a future use different than its original use. This avoids the environmental impact of using new materials.

air quality standards the level of pollutants prescribed by regulations that are not to be exceeded during a given time in a defined area. (EPA)

albedo the reflectivity of a surface, measured from 0 (black) to 1 (white).

alternative fuel vehicle a venhicle that uses low-polluting, non-gasoline fuels, such as electricity, hydrogen, propane or compressed natural gas, liquid natural gas, methanol, and ethanol.

ambient temperature the temperature of the surrounding air or other medium. (EPA)

ASHRAE American Society of Heating, Refrigerating and Air-Conditioning Engineers.

bake-out a process used to remove volatile organic compounds (VOCs) from a building by elevating the temperature in the fully furnished and ventilated building prior to human occupancy.

baseline versus design or actual use the amount of water that the design case or actual usage (for existing building projects) conserves over the baseline case. All Water Efficiency credits use a baseline case against which the facility's design case or actual use is compared. The baseline case represents the Energy Policy Act of 1992 (EPAct 1992) flow and flush rates and the design case is the water anticipated to be used in the facility.

biodegradable capable of decomposing under natural conditions. (EPA)

biodiversity the variety of life in all forms, levels, and combinations, including ecosystem diversity, species diversity, and genetic diversity.

biomass plant material from trees, grasses, or crops that can be converted to heat energy to produce electricity.

bioswale a stormwater control feature that uses a combination of an engineered basin, soils, and vegetation to slow and detain stormwater, increase groundwater recharge, and reduce peak stormwater runoff.

blackwater wastewater from toilets and urinals; definitions vary under some state or local codes and blackwater may include wastewater from kitchen sinks (perhaps differentiated by the use of a garbage disposal), showers, or bathtubs.

British thermal unit (Btu) the amount of heat required to raise the temperature of one pound of liquid water from 60° to 61° Fahrenheit. This standard measure of energy is used to describe the energy content of fuels and compare energy use.

brownfield previously used or developed land that may be contaminated with hazardous waste or pollution. Once any environmental damage has been remediated, the land can be reused. Redevelopment on brownfields provides an important opportunity to restore degraded urban land while promoting infill and reducing sprawl.

building commissioning verification after construction that the structure and its systems and subsystems meet project requirements as intended and designed.

building density the floor area of the building divided by the total area of the site (square feet per acre).

building envelope the exterior surface of a building—the walls, windows, roof, and floor; also referred to as the building shell.

building footprint the area on a project site that is used by the building structure, defined by the perimeter of the building plan. Parking lots, landscapes, and other non-building facilities are not included in the building footprint.

built environment any environment that is man made and provides a stucture for human activity.

byproduct material, other than the principal product, generated as a consequence of an industrial process or as a breakdown product in a living system. (EPA)

carbon dioxide concentrations an indicator of ventilation effectiveness inside buildings. CO2 concentrations greater than 530 parts per million (ppm) above outdoor conditions generally indicate inadequate ventilation. Absolute concentrations of greater than 800 to 1,000 ppm generally indicate poor air quality for breathing. CO2 builds up in a space when there is not enough ventilation.

carbon footprint a measure of greenhouse gas emissions associated with an activity. A comprehensive carbon footprint includes building construction, operation, energy use, building-related transportation, and the embodied energy of water, solid waste, and construction materials.

carbon neutrality projects that emit no more carbon emissions than they can either sequester or offset.

charrette intense workshops designed to produce specific deliverables.

chiller a device that removes heat from a liquid, typically as part of a refrigeration system used to cool and dehumidify buildings.

chlorofluorocarbon (CFC) an organic chemical compound known to have ozone-depleting potential.

clerestory windows located on interior rise above adjacent rooftops designed to admit daylight to interior spaces.

closed system a system that exchanges minimal materials and elements with its surroundings; systems are linked with one another to make the best use of byproducts.

commissioning (Cx) the process of verifying and documenting that a building and all of its systems and assemblies are planned, designed, installed, tested, operated, and maintained to meet the owner's project requirements.

commissioning plan a document that outlines the organization, schedule, allocation of resources, and documentation requirements of the commissioning process.

commissioning report a document that details the commissioning process, including a commissioning program overview, identification of the commissioning team, and description of the commissioning process activities.

community connectivity the amount of connection between a site and the surrounding community, measured by proximity of the site to homes, schools, parks, stores, restaurants, medical facilities, and other services and amenities. Connectivity benefits include more satisfied site users and a reduction in travel associated with reaching services.

compact fluorescent lamp (CFL) a small fluorescent lamp, used as a more efficient alternative to incandescent lighting; also called a PL, twin-tube, or biax lamp. (EPA)

construction and demolition debris waste and recyclables generated from construction and from the renovation, demolition, or deconstruction of preexisting structures. It does not include land-clearing debris, such as soil, vegetation, and rocks.

construction waste management plan a plan that diverts construction debris from landfills through recycling, salvaging, and reusing.

contaminant an unwanted airborne element that may reduce indoor air quality (ASHRAE Standard 62.1–2007).

controllability of systems the percentage of occupants who have direct control over temperature, airflow, and lighting in their spaces.

cooling tower a structure that uses water to absorb heat from air-conditioning systems and regulate air temperature in a facility.

cradle to cradle an approach in which all things are applied to a new use at the end of a useful life.

cradle to grave a linear set of processes that lead to the ultimate disposal of materials at the end of a useful life.

daylighting the controlled admission of natural light into a space, used to reduce or eliminate electric lighting.

development density the total square footage of all buildings within a particular area, measured in square feet per acre or units per acre.

diversion rate the percentage of waste materials diverted from traditional disposal methods to be recycled, composted, or re-used.

diversity of uses or housing types the number of types of spaces or housing types per acre. A neighborhood that includes a diversity of uses—offices, homes, schools, parks, stores—encourages walking, and its residents and visitors are less dependent on personal vehicles. A diversity of housing types allows households of different types, sizes, ages, and incomes to live in the same neighborhood.

dry ponds excavated areas that detain stormwater and slow runoff but are dry between rain events. Wet ponds serve a similar function but are designed to hold water all the time.

ecosystem a basic unit of nature that includes a community of organisms and their nonliving environment linked by biological, chemical and physical process.

embodied energy the total amount of energy used to harvest or extract, manufacture, transport, install and use a product across its life cycle.

emergent properties patterns that emerge from a system as a whole, which are more than the sum of the parts.

energy efficiency using less energy to accomplish the same amount of work.

energy management system a control system capable of monitoring environmental and system loads and adjusting HVAC operations accordingly in order to conserve energy while maintaining comfort. (EPA)

ENERGY STAR Portfolio Manager an interactive, online management tool that supports tracking and assessment of energy and water consumption.

ENERGY STAR rating a measure of a building's energy performance compared with that of similar buildings, as determined by the ENERGY STAR Portfolio Manager. A score of 50 represents average building performance.

energy use intensity energy consumption divided by the number of square feet in a building, often expressed as British thermal units (Btus) per square foot or as kilowatt-hours of electricity per square foot per year (kWh/sf/yr).

energy-efficient products and systems building components and appliances that use less energy to perform as well as or better than standard products.

environmental sustainability long-term maintenance of ecosystem components and functions for future generations. (EPA)

externality costs or benefits, separate from prices, resulting from a transaction and incurred by parties not involved in the transaction.

feedback loop information flows within a system that allow the system to self-organize.

floodplain land that is likely to be flooded by a storm of a given size (e.g., a 100-year storm).

floor-area ratio the relationship between the total building floor area and the allowable land area the building can cover. In green building, the objective is to build up rather than out because a smaller footprint means less disruption of the existing or created landscape.

flush-out the operation of mechanical systems for a minimum of two weeks using 100 percent outside air at the end of construction and prior to building occupancy to ensure safe indoor air quality.

footcandle a measure of the amount of illumination falling on a surface. A footcandle is equal to one lumen per square foot. Minimizing the number of footcandles of site lighting helps reduce light pollution and protect dark skies and nocturnal animals.

fossil fuel energy derived from ancient organic remains, such as peat, coal, crude oil, and natural gas. (EPA)

gallons per flush (gpf) the amount of water consumed by flush fixtures (water closets, or toilets, and urinals).

gallons per minute (gpm) the amount of water consumed by flow fixtures (lavatory faucets, showerheads, aerators).

green building encompasses planning, design, construction, operations, and ultimately end-of-life recycling or renewal of structures. Green building pursues solutions that represent a healthy and dynamic balance between environmental, social, and economic benefits.

green power energy from renewable sources such as solar, wind, wave, biomass, geothermal power and several forms of hydroelectric power.

greenfield a site that has never been developed for anything except agriculture.

greenwashing presenting misinformation to consumers to portray a product or policy as more environmentally friendly than it actually is.

greywater domestic wastewater composed of wash water from kitchen, bathroom, and laundry sinks, tubs, and washers which has not come into contact with toilet waste. Some states and local authorities allow kitchen sink wastewater to be included in greywater.

harvested rainwater precipitation captured and used for indoor needs, irrigation, or both.

heat island effect the absorption of heat by hardscapes, such as dark, nonreflective pavement and buildings, and its radiation to surrounding areas. Particularly in urban areas, other sources may include vehicle exhaust, air-conditioners, and street equipment; reduced airflow from tall buildings and narrow streets exacerbates the effect.

high-performance green building a structure designed to conserve water and energy; use space, materials, and resources efficiently; minimize construction waste; and create a healthful indoor environment.

hydrochlorofluorocarbons (HCFC) an organic chemical compound known to have ozone-depleting potential.

HVAC systems equipment, distribution systems, and terminals that provide the processes of heating, ventilating, or air-conditioning. (ASHRAE Standard 90.1–2007)

impervious area surface that has been compacted or covered by materials that do not allow water to infiltrate. Impervious areas found in the built environment include concrete, brick, stone, asphalt, and sealed surfaces.

imperviousness the resistance of a material to penetration by a liquid. The total imperviousness of a surface, such as paving, is expressed as a percentage of total land area that does not allow moisture penetration. Impervious surfaces prevent rainwater from infiltrating into the ground, thereby increasing runoff, reducing groundwater recharge, and degrading surface water quality.

indoor air quality the nature of air inside the space that affects the health and well-being of building occupants. It is considered acceptable when there are no known contaminants at harmful concentrations and a substantial majority (80% or more) of the occupants do not express dissatisfaction. (ASHRAE Standard 62.1–2007)

Indoor Air Quality Building Education and Assessment Model (I-BEAM) an integral part of an IAQ management program that provides comprehensive guidance for building professionals responsible for indoor air quality in commercial buildings. Incorporates an IAQ audit of the project building to determine the building's IAQ status.

indoor environmental quality the conditions inside a building and their impacts on occupants or residents.

indoor environmental quality management plan a plan that spells out strategies to protect the quality of indoor air for workers and occupants; it includes isolating work areas to prevent contamination of occupied spaces, timing construction activities to minimize exposure to off-gassing, protecting the HVAC system from dust, selecting materials with minimal levels of toxicity, and thoroughly ventilating the building before occupancy.

infill development is a method of site selection that focuses construction on sites that have been previously developed or are gaps between existing structures.

integrated design team all the individuals involved in a building project from early in the design process, including the design professionals, the owner's representatives, and the general contractor and subcontractors.

integrated pest management a sustainable approach that combines knowledge about pests, the environment, and pest prevention and control methods to minimize pest infestation and damage in an economical way while minimizing hazards to people, property, and the environment.

integrated process an approach to design and operations that brings team members together to work collaboratively on all of the project's systems to find synergistic solutions that support ever greater levels of sustainability.

irrigation efficiency the percentage of water delivered by irrigation equipment that is actually used for irrigation and does not evaporate, blow away, or fall on hardscape. For example, overhead spray sprinklers have lower irrigation efficiencies (65%) than drip systems (90%).

iterative process circular and repetitive process that provides opportunities for setting goals and checking each idea against those goals

LEED credit an optional component of the LEED rating system whose achievement results in the earning of points toward certification.

LEED Credit Interpretation Request a formal USGBC process in which a project team experiencing difficulties in the application of a LEED prerequisite or credit can seek and receive clarification, issued as a credit interpretation ruling. Typically, difficulties arise when specific issues are not directly addressed by LEED reference guides or a conflict between credit requirements arises.

LEED intent the primary goal of each prerequisite or credit.

LEED Interpretation potentially precedent-setting responses to Credit Interpretation Requests made by project teams.

LEED Minimum Program Requirements (MPRs) a list of the basic characteristics that a project must possess to be eligible for certification therefore defining a broad category of buildings that the LEED rating systems were designed to evaluate.

LEED Online a data collection portal managed by GBCI through which the team uploads information about the project.

LEED Pilot Credit Library credits currently being tested across rating systems and credit categories that are proposed for the next version of LEED.

LEED prerequisite a required LEED Green Building Rating System component whose achievement is mandatory and does not earn any points.

LEED Rating System a voluntary, consensus-based, market-driven building rating system based on existing, proven technology. The LEED Green Building Rating System represents USGBC's effort to provide a national benchmark for green buildings. Through its use as a design guideline and third-party certification tool, the LEED Green Building Rating System aims to improve occupant well-being, environmental performance, and economic returns using established and innovative practices, standards, and technologies.

LEED technical advisory group (TAG) a committee consisting of industry experts who assist in interpreting credits and developing technical improvements to the LEED Green Building Rating System.

leverage point a point in a system where a small intervention can yield large changes.

life cycle approach looking at all stages of a project, product or service, adding the dimension of longevity to whole systems thinking

life cycle assessment an analysis of the environmental aspects and potential impacts associated with a product, process, or service.

life cycle costing a process of costing that looks at both purchase and operating costs as well as relative savings over the life of the building or product

light trespass the spillage of light beyond the project boundary.

lighting power density the installed lighting power per unit area.

low impact development (LID) an approach to land management that mimics natural systems to manage stormwater as close to the source as possible.

market transformation systematic improvements in the performance of a market or market segment.

materials reuse materials returned to active use (in the same or a related capacity as their original use), expressed as a percentage of the total materials cost of a building. The salvaged materials are incorporated into the new building, thereby extending the lifetime of materials that would otherwise be discarded.

measures of energy use typical primary measures of energy consumption associated with buildings include kilowatt-hours of electricity, therms of natural gas, and gallons of liquid fuel.

Minimum Efficiency Reporting Value (MERV) a rating that indicates the efficiency of air filters in the mechanical system. MERV ratings range from 1 (very low efficiency) to 16 (very high).

Montreal Protocol an international treaty that eliminates or partially eliminates the use of substances known to deplete the ozone layer.

native (or indigenous) plants plants adapted to a given area during a defined time period. In North America, the term often refers to plants growing in a region prior to the time of settlement by people of European descent. Native plants are considered low maintenance and not invasive.

negative feedback loop a signal for a system to stop changing when a response is no longer needed.

net-zero energy projects that use no more energy from the grid than they can produce on site.

nonpoint source pollution typically refers to water pollution caused by stormwater runoff from diffuse sources. When it rains, water washes fertilizers, car oil, pet waste, etc, into receiving water bodies.

nonpotable water See potable water.

nonrenewable not capable of being replaced; permanently depleted once used. Examples of nonrenewable energy sources are oil or natural gas, and nonrenewable natural resources include metallic ores.

occupant comfort survey measures occupant comfort level in a variety of ways, including thermal comfort, acoustics, indoor air quality, lighting levels, and building cleanliness.

off-gassing the emission of volatile organic compounds from synthetic and natural products.

open system a system in which materials are constantly brought in from the outside, used in the system, and then released outside the system in a form of waste.

particulates solid particles or liquid droplets in the atmosphere. The chemical composition of particulates varies, depending on location and time of year. Sources include dust, emissions from industrial processes, combustion products from the burning of wood and coal, combustion products associated with motor vehicle or nonroad engine exhausts, and reactions to gases in the atmosphere. (EPA)

passive design planning with the intent of capturing sunlight, wind or other natural forces for light, heating, and cooling.

performance monitoring continously tracking efficiency metrics of energy, water and other systems, specifically to respond and achiever better levels of efficiency.

performance relative to benchmark a comparison of the performance of a building system with a standard, such as ENERGY STAR Portfolio Manager.

performance relative to code a comparison of the performance of a building system with a baseline equivalent to minimal compliance with an applicable energy code, such as ASHRAE Standard 90.1 or California's Title 24.

perviousness the percentage of the surface area of a paving material that is open and allows moisture to pass through the material and soak into the ground below.

pest control management a sustainable approach that combines knowledge about pests, the environment, and pest prevention and control methods to minimize pest infestation and damage in an economical way while minimizing hazards to people, property, and the environment.

photovoltaic (PV) energy electricity from photovoltaic cells that convert the energy in sunlight into electricity.

pollutant any substance introduced into the environment that adversely affects the usefulness of a resource or the health of humans, animals, or ecosystems. (EPA) Air pollutants include emissions of carbon dioxide (CO_2), sulfur dioxide (SO_2), nitrogen oxides (NO_x), mercury (Hg), small particulates (PM2.5), and large particulates (PM10).

positive feedback loop self-reinforcing loops in which a stimulus causes an effect and the loop produces more of that effect.

postconsumer recycled content the percentage of material in a product that was consumer waste. The recycled material was generated by household, commercial, industrial, or institutional end users and can no longer be used for its intended purpose. It includes returns of materials from the distribution chain. Examples include construction and demolition debris, materials collected through recycling programs, discarded products (e.g., furniture, cabinetry, decking), and landscaping waste (e.g., leaves, grass clippings, tree trimmings). (ISO 14021)

potable water is water meeting quality standard allowing for consumption or use with low risk of immediate or long term harm.

preconsumer recycled content the percentage of material in a product that was recycled from manufacturing waste. Preconsumer content was formerly known as postindustrial content. Examples include planer shavings, sawdust, bagasse, walnut shells, culls, trimmed materials, overissue publications, and obsolete inventories. Excluded are rework, regrind, or scrap materials capable of being reclaimed within the same process that generated them. (ISO 14021)

prime farmland previously undeveloped land with soil suitable for cultivation. Avoiding development on prime farmland helps protect agricultural lands, which are needed for food production.

project administrator the individual from the project team that registers a project with GBCI.

Project credit interpretation rulings (CIR) a response from GBCI providing technical guidance on how LEED requirements pertain to particular projects.

project team a broad, inclusive, collaborative group that works together to design and complete a project.

rain garden a stormwater management feature consisting of an excavated depression and vegetation that collect and infiltrate runoff and reduce peak discharge rates.

rainwater harvesting the collection and storage of precipitation from a catchment area, such as a roof.

rapidly renewable materials agricultural products (fiber or animal) that are grown or raised quickly and can be harvested in a sustainable fashion, expressed as a percentage of the total materials cost. For LEED, rapidly renewable materials take 10 years or less to grow or raise.

recycled content the percentage of material in a product that is recycled from the manufacturing waste stream (preconsumer waste) or the consumer waste stream (postconsumer waste) and used to make new materials. For LEED, recycled content is typically expressed as a percentage of the total material volume or weight.

refrigerant one of any number of substances used in cooling systems to transfer thermal energy in air conditioning and refrigeration systems.

regenerative evolving with living systems and contributing to the long term renewal of resources and the health of all life in each unique place

regenerative design sustainable plans for built environments that improve existing conditions. Regenerative design goes beyond reducing impacts to create positive change in the local and global environment.

regional materials materials that are extracted, processed, and manufactured close to a project site, expressed as a percentage of the total materials cost. For LEED, regional materials originate within 500 miles of the project site.

renewable energy resources that are not depleted by use. Examples include energy from the sun, wind, and small (low-impact) hydropower, plus geothermal energy and wave and tidal systems. Ways to capture energy from the sun include photovoltaic, solar thermal, and bioenergy systems based on wood waste, agricultural crops or residue, animal and other organic waste, or landfill gas.

renewable energy certificate (REC) a tradable commodity representing proof that a unit of electricity was generated from a renewable energy resource. RECs are sold separately from the electricity itself and thus allow the purchase of green power by a user of conventionally generated electricity.

retrocommissioning a commissioning process that can be performed on existing buildings to identify and recognize system improvements that make the building more suitable for current use.

salvaged material construction items recovered from existing buildings or construction sites and reused. Common salvaged materials include structural beams and posts, flooring, doors, cabinetry, brick, and decorative items.

sick building syndrome (SBS) a combination of symptoms, experienced by occupants of a building, that appear to be linked to time spent in the building but cannot be traced to a specific cause. Complaints may be localized in a particular room or zone or be spread throughout the building. (EPA)

site disturbance the amount of a site that is disturbed by construction activity. On undeveloped sites, limiting the amount and boundary of site disturbance can protect surrounding habitat.

smart growth an approach to growth that protects open space and farmland by emphasizing development with housing and transportation choices near jobs, shops and schools.

solar reflectivity index (SRI) a measure of how well a material rejects solar heat; the index ranges from 0 (least reflective) to 100 (most reflective). Using light-colored, "cooler" materials helps prevent the urban heat island effect (the absorption of heat by dark roofs and pavement and its radiation to the ambient air) and minimizes demand for cooling of nearby buildings.

stakeholder a dynamic term that encompasses a broad array of individuals tasked with the design, creation, and operation of a building as well as those whose lives will be impacted by the built environment at hand.

stakeholder meetings meetings that include those with a vested interest in the outcome of a project.

stormwater prevention plan a plan that addresses measures to prevent erosion, sedimentation, and discharges of potential pollutants to water bodies and wetlands.

stormwater runoff water from precipitation that flows over surfaces into sewer systems or receiving water bodies. All precipitation that leaves project site boundaries on the surface is considered stormwater runoff.

street grid density an indicator of neighborhood density, calculated as the number of centerline miles per square mile. Centerline miles are the length of a road down its center. A community with high street grid density and narrow, interconnected streets is more likely to be pedestrian friendly than one with a low street grid density and wide streets.

sustainability meeting the needs of the present without compromising the ability of future generations to meet their own needs. (Brundtland Commission)

sustainable forestry the practice of managing forest resources to meet the long-term forest product needs of humans while maintaining the biodiversity of forested landscapes.

system an assemblage of parts that interact in a series of relationships to form a complex whole, which serves particular functions of purposes.

systems thinking understanding the world, including the built environment, as a series of relationships in which all parts influence many other parts.

thermal comfort the temperature, humidity, and airflow ranges within which the majority of people are most comfortable, as determined by ASHRAE Standard 55. Because people dress differently depending on the season, thermal comfort levels vary with the season. Control setpoints for HVAC systems should vary accordingly to ensure that occupants are comfortable and energy is conserved.

transportation demand management the process of reducing peak-period vehicle trips.

triple bottom line incorporates a long-term view for assessing potential effects and best practices for three kinds of resources: people, planet, profit.

value engineering a formal review process of the design of a project based on its intended function in order to identify potential alternatives that reduce costs and improve performance.

vehicle miles traveled (vmt) a measure of transportation demand that estimates the travel miles associated with a project, most often for single-passenger cars. LEED sometimes uses a complementary metric for alternative-mode miles (e.g., in high-occupancy autos).

ventilation rate the amount of air circulated through a space, measured in air changes per hour (the quantity of infiltration air in cubic feet per minute divided by the volume of the room). Proper ventilation rates, as prescribed by ASHRAE Standard 62, ensure that enough air is supplied for the number of occupants to prevent accumulation of carbon dioxide and other pollutants in the space.

volatile organic compound (VOC) substances that vaporize at room temperature and can cause health problems. VOCs off-gas from many materials, including adhesives, sealants, paints, carpets, and particle board. Limiting VOC concentrations protects the health of both construction personnel and building.

waste diversion the amount of waste disposed other than through incineration or in landfills, expressed in tons. Examples of waste diversion include reuse and recycling.

waste management plan a plan that addresses the sorting, collection, and disposal of waste generated during construction or renovation. It must address management of landfill waste as well as recyclable materials.

wastewater the spent or used water from a home, community, farm, or industry that contains dissolved or suspended matter. (EPA)

water balance a comparative measure of the amount of water that flows in and out of a system.

wetland vegetation plants that require saturated soils to survive or can tolerate prolonged wet soil conditions.

Wingspread Principles on a U.S. Response to Global Warming a set of propositions signed by individuals and organizations declaring their commitment to addressing the issue of climate change.

xeriscaping a landscaping method that makes routine irrigation unnecessary by using drought-adaptable and low-water plants, as well as soil amendments such as compost and mulches to reduce evaporation.

RECYCLED
Paper made from
recycled material
FSC® C014174

This reference guide was printed on 100% postconsumer waste paper, processed chlorine free, and printed with non-toxic, soybased inks using 100% wind power. By using these materials and production processes, the U.S. Green Building Council saved the following resources:

Trees*	Solid Waste	Liquid Waste	Electricity	Greenhouse Gases	Sulfur Dioxides	Nitrogen Oxides
1259 trees preserved for the future	59,161 lbs.	534,707 gallons	891,616,000 BTUs	116,487 lb.	N/A	N/A

*one harvested tree = aprox. 575 lbs

RCV 11:19

ISBN 13: 978-0-13-291509-0
ISBN 10: 0-13-291509-X

90000>

9 780132 915090

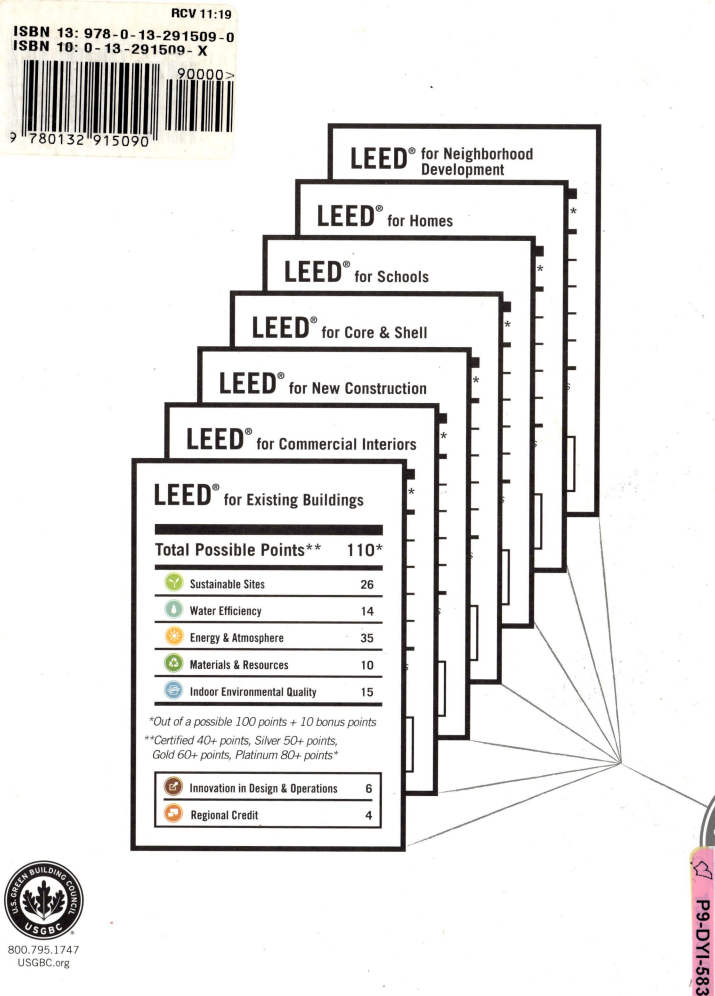

LEED® for Neighborhood Development

LEED® for Homes

LEED® for Schools

LEED® for Core & Shell

LEED® for New Construction

LEED® for Commercial Interiors

LEED® for Existing Buildings

Total Possible Points**	110*
Sustainable Sites	26
Water Efficiency	14
Energy & Atmosphere	35
Materials & Resources	10
Indoor Environmental Quality	15

*Out of a possible 100 points + 10 bonus points

**Certified 40+ points, Silver 50+ points,
Gold 60+ points, Platinum 80+ points*

Innovation in Design & Operations	6
Regional Credit	4

U.S. GREEN BUILDING COUNCIL
USGBC®

800.795.1747
USGBC.org

P9-DYI-583